OPTICS

An Introduction for
Students of Engineering

OPTICS

An Introduction for Students of Engineering

J. Warren Blaker
Fairleigh Dickinson University

William M. Rosenblum
University of Alabama at Birmingham

Macmillan Publishing Company
New York

Maxwell Macmillan Canada
Toronto

Maxwell Macmillan International
New York   Oxford   Singapore   Sidney

Editor: Robert McConnin
Production Supervisor: Margaret Comaskey
Production Manager: Roger Vergnes
Text Designer: Natasha Sylvester
Cover Designer: *Natasha Sylvester*

This book was set in Times Roman and ITC Lubalin Graph Book by York Graphic Services, printed and bound by Book Press. The cover was printed by Lehigh.

Macmillan Publishing Company is part
of the Maxwell Communication Group of Companies.

Macmillan Publishing Company
866 Third Avenue, New York, New York 10022

Maxwell Macmillan Canada, Inc.
1200 Eglinton Avenue East
Suite 200
Don Mills, Ontario M3C 3N1

Library of Congress Cataloging in Publication Data

Library of Congress Cataloging-in-Publication Data

Blaker, J. Warren.
    Optics—an introduction for students of engineering/J. Warren
Blaker, William M. Rosenblum.
        p.   cm.
    Includes index.
    ISBN 0-02-310640-9
    1. Optics.   I. Rosenblum, William M.   II. Title.
QC355.2.B53   1993
535—dc20   92-18269
                                                                 CIP

Printing: 1 2 3 4 5 6 7 8   Year: 3 4 5 6 7 8 9 0 1 2

PREFACE

This text is designed to meet the needs of the engineering student. Many shifts in technology, including the introduction of new fiber optical communications systems, have increased the importance of a basic understanding of optics for all engineers. The typical optics text has in the past been directed to an undergraduate physics student and has emphasized features and applications which served as foundations for further study in physics. The engineering student, in contrast, needs, in addition to a grounding in the models used in optics, an overview of design and a discussion of engineering applications. This has been our goal.

Optics is model based. By this we mean that the detailed understanding of optical phenomena can be developed via computational models. Here we have used the traditional ray and wave (physical) models to develop optics. We have introduced some of the design elements so important to an engineering education and have treated as applications such things as fibers, lasers, and holography.

Following a short historical introduction in Chapter 1, the ray model is introduced in Chapter 2 using the laws of reflection and refraction as a basis. This chapter introduces imaging using the common forms of the lens and mirror equations and presents, as a contrast, Newton's form of the lens equation. We have emphasized the importance of diagramming systems, not only as a check on calculation, but also in order to give an insight into the behavior of the system.

The matrix form for treating optical systems is presented in Chapter 3 to provide the tools for treating more complex systems in a uniform way. This formulation also provides an opportunity to introduce the cardinal points of the system in a coherent manner. There are two ways in which these matrices can be written; we have used the one which puts the negative signs in the matrix and makes the signs in the object–image equation all positive.

Chapter 4 is an introduction to optical design. Here the aberrations of the optical system are introduced as part of a discussion of the design process. The discussion of design presents many of the concerns facing the optical designer but it is intended to be an example rather than an exhaustive treatment of design. Chapter 5 follows with a number of examples of optical systems commonly found in optical instruments. The selection is somewhat arbitrary but this can easily be extended if the instructor chooses to do so.

Chapter 6 introduces the sections of the book devoted to the optical wave model. This chapter shows how the wave model arises from Maxwell's equations and introduces interferometry to illustrate the experimental foundation of the wave model.

Chapters 7 and 8 extend the wave model through application of interferometry and an introduction to diffraction effects. Both these areas are important to engineers in metrology and in the understanding of antennas.

The final four chapters treat applications. Chapter 9 discusses light sources with an emphasis on lasers. Chapter 10 treats fiber optical systems, which are so important to communications engineering today. Chapter 11 shows how the ideas of linear system theory apply to optics and continues with a brief discussion of optical computing. Finally, Chapter 12 deals with image recording, treating both film and electromagnetic recording. This final chapter also introduces holography.

Four appendixes are included. Those treating matrices and Fourier methods are to serve as mathematical refreshers for the students. Two others deal with the derivation of the fundamental diffraction laws and with the treatment of waves in material media. The latter two can be used by the instructor to augment the treatment in the text or simply read by the student for background information.

The authors want to thank Avi, Judy, and Vasha for their patience during the period in which this book was being developed. Often family time was sacrificed in order to complete this effort and we hope the result will be worth the sacrifice.

In addition, we want to thank those students who used various portions of the text in class over a number of years, as well as the reviewers. They are due thanks for any number of valuable suggestions which we believe have made the book better. Any errors, however, remain ours.

Finally we want to thank Bob McConnin for his friendship and support over many years and Margaret Comaskey and her production group for their help and patience.

CONTENTS

CHAPTER 1

Introduction

The study of optics extends well beyond the usual intuitive concept of the understanding and manipulation of light. Optics really involves the application of various *computational models* to the electromagnetic spectrum in the region ranging from x rays to microwaves and occasionally beyond these rather vague limits. Thus this text will be devoted to the presentation, development, and use of optical computational models over a rather wide spectral range, but the examples will be drawn principally from the visible spectrum where the reader will have some intuitive understanding.

Optics began with the contemplation of the eye as the detector of light. As with any information transmitting system, one needs a source, a transmitting channel, and a detector. The eye was the detector of light transmitted through the atmosphere from astronomical sources or from fire. The limits of the optics in this instance are set by the eye's sensitivity, that is, the electromagnetic wavelengths between 400 and 700 nm (1 nm = 10^{-9} m). This spectral range is a very tiny portion of the electromagnetic spectrum, albeit an important one. The study of the optics of vision belongs to a branch of optics known as *physiological optics*.

The manipulation of light with lenses, prisms, mirrors, and such devices based on the model of rectilinear propagation of electromagnetic radiation is called *geometrical optics*. Geometrical optics is the oldest of the computational models and began in earnest with the work of Euclid.

A deeper understanding of optical phenomena is often achieved, however, by looking to the nature of electromagnetic radiation itself. The study of *physical optics* involves examination of the propagation of optical waves.

The most modern quantum mechanical view treats light as ''wave packets'' which have properties closely resembling those which might be attributed to particles. Such ''light particles'' are called *photons,* and these form the basis for the model called *quantum optics.*

This text will treat the geometrical and wave models of light but will not delve very deeply into the quantum nature of light, since that would require a much deeper background in quantum physics than is expected of the readers.

The History of Optics

It is not possible to give anything but an extremely brief synopsis of a few relevant happenings in the history of optics here, but it is important that one understand that what follows in this text has developed over many centuries. The models which will be important in our approach to optics will be seen to have had their bases in the historical development of the field and certainly are not modern in origin.

Early humans clearly observed their images as reflected from the still water in a spring or lake. Indeed, the ancients made mirrors of polished metal, usually bronze, and there are ample references to mirrors in ancient writings. Lenses, or at least one object which is probably a lens, appear to have dated from at least 4000 years ago. A piece of cut and polished quartz in the crude form of a lens was found with a cuneiform tablet with tiny inscription during excavation of Sumerian ruins in Mesopotamia. While no known writings describe it, its discovery in juxtaposition with the tablet leaves little doubt as to its function.

The early Greek philosophers devoted considerable effort to the understanding of vision, and they thought of light as both rays and particles. Writings, including those of both Plato and Aristotle, that date from about 2500 years ago are among the earliest known dealing with optical phenomena and primarily represent the thoughts of the philosophers on the nature of the visual process. A century later Euclid produced the first optics texts, *Optics* and *Catoptics,* dealing with vision in the first and reflection in the latter. Euclid is believed to have been the first to state the law of reflection correctly. Euclid with his obvious geometrical interests extended the ideas of mirrors to spherical surfaces, defined the focal point, and examined virtual images.

Archimedes is widely reported by early historians to have used mirrors to set fire to the Roman fleet at Syracuse. It seems likely that Archimedes was able to focus the sun's light with a mirror and thereby kindle a fire, but it seems improbable that he was able to set fire to the entire Roman fleet.

While it would seem that the early Greeks were deeply involved in the technical aspects of optics, in reality their concern was largely philosophical and dealt with issues such as ''how do we see.'' They also tried to understand color and its origin. These issues were of such importance to them that each of the philosophical schools seems to have developed its own theory. The debates must have been numerous. Plato's *Dialogues* contain many references to the ideas put forward at that time.

One of the great mathematicians of the Greek era, Hero of Alexandria, wrote an optics monograph entitled *Catoptics*. In this book he developed the concept that the straight-line propagation of light was due to a minimum path principle. In this he was a forerunner to Fermat, who expanded this idea some 1600 years later in 1658 to the more general extremum principle that bears his name.

During the first millennium of this era, little advancement in the understanding of optics was recorded. With the rise of scientific thought in the East, Alhazen of Cairo, best known for his contributions to mathematics, wrote a text entitled *Optics* in about the year 1100. This monumental work not only accurately discussed the anatomy of the eye but, equally importantly, dealt with refraction. Alhazen knew that a glass sphere magnified images and he discussed the pinhole camera or camera obscura in some detail. In addition, Alhazen discussed the origin of the rainbow and imaging errors, specifically, spherical aberration. The work of Alhazen and Euclid's works are the foundations of optics as a scientific discipline and they delineated the early geometrical model.

The contributions of Alhazen reached Europe through the Moorish conquest of Spain and contributed significantly to the understanding of optics that developed during the Renaissance. From the 10th to the 16th centuries, there was little activity in the science of optics. However, spectacles are first seen in a painting of Hugh of Provence by Tommaso da Modena dated 1352 but may have originated in the Orient. In the writings of Marco Polo, for example, one sees that exchange between the East and West was taking place at this time. In the 15th century Leonardo da Vinci examined the pinhole camera and again considered the origin and nature of color as part of his vast researches.

Bacon was aware of the work of Alhazen, and optics was the only area in which he experimented actively. In fact, he described a telescope as well as lenses and mirrors.

With the 16th century came Kepler, who was influenced by Alhazen's writings, which had been rendered into Latin, as well as by the writings of the Greek philosophers. While he is best known for his laws of planetary motion, in a book published in 1604 he extended the ideas of optical astronomy and provided the foundation for the understanding of the functioning of the eye by correctly establishing the function of the retina. Kepler also considered the problem of visual defects, which he attributed to changes in the length of the eye; the language he introduced in calling nearsightedness a long eye and farsightedness a short eye persists today. Equally important in Kepler's work was his ideas about extended images, where he envisioned an extended object as a collection of point sources and images as the collection of point images. This model is still important and will be treated later in this text in the discussion of linear systems.

Galileo Galilei (1564–1642) was a leader among the "modern" scientists in abandoning the purely philosophical approach of the Greeks for a truly experimentally based science. His name is often associated with the telescope, although this instrument was most certainly invented in Holland, but it was Galileo who recognized its importance. Galileo discovered the moons of Jupiter using his hand-held telescope. Galileo also tried to measure the velocity of light but was able to conclude only that the velocity was very large. Romer in 1675 was able to find a value approaching that accepted today by measuring the transit time of Jupiter's moons.

The law of refraction had been discovered by Snell in 1621 and possibly independently by Descartes at about the same time. Fermat showed that this was consistent with a least-time principle later extended to a least-action principle by Maupertuis, reviving the concepts of Hero of Alexandria.

The 17th century was a time of great change in science, dominated by the figure of Newton (1642–1727). Newton's first investigations seem to have been devoted to color and the nature of dispersion, and this as much as anything marked the end of the purely geometrical view of the nature of light. Newton not only refracted light with a prism and demonstrated the colored components, he also experimented with the contact between a flat prism surface and a lens and observed the interference effect still known as Newton's rings.

Newton also studied the double refraction from Iceland spar and thus had considerable experimental data pointing to a wave theory for light. Nonetheless, throughout his lifetime he held that light was particulate in nature and not a wave. A number of scientists of Newton's epoch, Malebranche, Hooke, Grimaldi, and Huygens among them, pressed strongly for a vibratory or wave theory. A great scientific debate on the question of the nature of light began.

Both Hooke and Grimaldi had discovered the diffraction in the shadow of a straightedge. Newton not only repeated this experiment but extended it to a slit aperture where he was able to see the diffraction effect more strongly. The arguments regarding the nature of light would rage until the beginning of the 19th century.

The wave theory of light received an impetus and was placed on a firm footing through the work of Fresnel and Thomas Young. Young observed the light transmitted by two pinholes in a screen illuminated by a narrow beam. On a second screen he observed the bright interference bands due to the superposition of the light waves. This apparatus allowed him to calculate the very small wavelength of light.

Fresnel developed a mathematical theory of light to notable degree. Based on his theoretical development he predicted the bright central spot in diffraction about a circular disk. Although his result was disputed by the French Academy for some time, his calculation was a major success in the theoretical understanding of light.

The problem of a medium which would sustain the propagation of light waves remained an issue. However, beginning in 1855, James Clerk Maxwell developed his theory of electricity and magnetism. One important consequence of this theory was the establishment of light as an electromagnetic wave, based on the identification of the propagation velocity of electromagnetic waves with that of light. This received further verification when Hertz experimentally observed the nonoptical electromagnetic waves. An additional problem arose with the discovery of the photoelectric effect by Hertz in 1887. The photoelectric effect contains experimental findings fundamentally inconsistent with the wave theory. Albert Einstein was able to explain the photoelectric effect on the basis of a particle description of light. The light particles are called photons and can be described through the use of quantum theory.

The problem of a propagation medium was approached by Michelson and Morley in their classical interferometry experiment. They were able to establish that the velocity of light is always the same, independent of the motion of the observer. This result implies that there is no need for a medium to support electromagnetic radiation. The Michelson–Morley experiment also laid part of the foundation of the theory of relativity.

Today quantum electrodynamics, the quantum theory of the electromagnetic field, represents the level of most detailed understanding of light. In this model light is treated as ''packets'' of energy known as *photons* with a defined wavelength but with no rest mass.

Computational Models

The historical material has shown that the view of the nature of light has changed from time to time. The particle streams or rays of the Greeks have given way to a wave picture, only to have that supplanted by photons in this century, a new particle picture. In spite of this evolution in the way that light and optics has been understood, significant progress has been made in nearly every era in the application of optics. Today's understanding of light as photons is very different from the ray picture used two centuries ago, and yet very effective optical instruments were designed in that era based on a ray model.

The progress which has occurred has resulted from the use of *physical computational models* which have allowed engineers and scientists to make calculations and design instruments based on an idealized picture of the behavior of light. Such models are certainly familiar to most students as weightless, frictionless pulleys and point masses from physics or other linear devices in engineering. In optics they take the form of rays, or waves, or photons. These models are *physical models* in that they are derived from observation of the behavior of light. A second kind of model, the *empirical model,* based on a statistical analysis of the data of a particular phenomenon as is used in the study of the mechanics of gases is different from the models used here.

The use of optical models is an extremely powerful tool as well as an effective way to learn about optics. This book is developed along the lines of these physical models, and separate sections of the book are devoted to the ray model and the wave model. One should carefully note the strength of each model as it is presented and consider particularly those cases in which either of these models can be used in understanding some problem. The quantum model is not treated here beyond mention in the sections dealing with lasers. For most engineering applications, the ray and wave models are more than satisfactory.

CHAPTER 2

The Geometrical Model

The first optical model to be treated here is the geometrical model. In this chapter the fundamental geometrical laws of reflection and refraction will be presented so that a first brief treatment of imaging can be developed. Emphasis will be given to establishing the sign conventions to be used in analyzing images as well as on spherical surfaces and their importance in imaging.

There are several choices available for use as the foundation for the geometrical model. Perhaps the most direct is to postulate the laws of reflection and refraction, and that is the path that will be followed here.

One idea used by the Greek philosophers in their discussions of light was the principle of rectilinear propagation. The idea of rectilinear propagation arises naturally if one examines the form of shadows. Shadows usually are geometrically similar to the object interrupting the path of light as it propagates from its source, as illustrated in Figure 2.1.

The similarity of T and T' in the figure suggests that the light leaving the source S travels in straight lines outward from S. Thus, based on shadowing, one can infer that light travels in straight lines, that is, by *rectilinear propagation,* at least in a uniform medium. In such cases the light can be represented by straight lines from the source, *light rays*. The tracing of the paths of light rays is the fundamental exercise in using the geometrical model.

The simplest example of the use of ray tracing arises during reflection.

The Law of Reflection. A light ray striking a reflecting surface and reflected from it makes equal angles with the normal to the surface at the point of incidence, and the incoming ray, the outgoing ray, and the normal to the surface at the point of intersection all lie in the same plane.

Figure 2.2 illustrates the law of reflection.

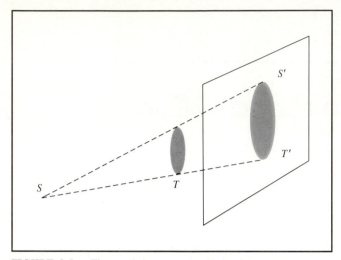

FIGURE 2.1. The point source of light S illuminates an opaque circular disk T and casts a circular shadow on the screen S'.

Reflection occurs at the interface between different media. For a mirror, for example, the reflecting surface is smooth and uniform, and the reflective process is termed *specular reflection*. The reflection from an irregular surface such as cloth or paper still obeys the law of reflection, but because of the irregularity of the surface, adjacent rays are reflected in highly divergent directions; such a process is called *diffuse reflection*.

Reflection always occurs whenever there is an abrupt change in the medium in which propagation is occurring. Consider a window as an example. When the outside light level is high, as in the daylight hours, the window transmits the outside illumination into the room and does not appear to function as a reflecting surface. It is only when the

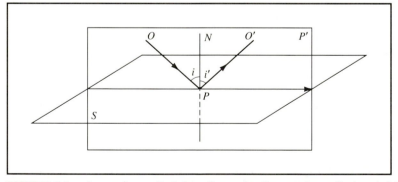

FIGURE 2.2. A ray from O strikes the reflecting surface S at point P. The incoming ray OP and the outgoing ray PO' make equal angles i and i' with the surface normal N in plane P.

outside light level falls well below that in the room that one sees that the window truly functions as a reflecting surface and the inside of the room can be seen to be reflected in the window. The window reflects approximately 5% of the light incident upon it, but in the presence of a bright scene outside the window, the reflected light is swamped by the light passing through the window from outside.

Images can be formed with specular (smooth) reflecting surfaces, the simplest example of which is a plane mirror; however, various nonplanar surfaces also form useful images. The spherical mirror will be examined later in this chapter.

EXAMPLE 2.1

How large a mirror is necessary if one wants to view one's full-length image?

Solution

Take the distance from your eyes to the mirror as l and the mirror as fixed vertically in Figure E2.1. By the congruence of the triangles EBM and FBM, the distances EB and BF are equal. Likewise, HE is divided into two equal parts. The mirror below M or above M' does not contribute to the person's image and the length MM' is precisely one-half the height of the person. Note that this mirror length is independent of l. One can see how the name *geometrical optics* for the ray model arose since this simple problem uses reasoning that was typical of early geometrical optics calculations.

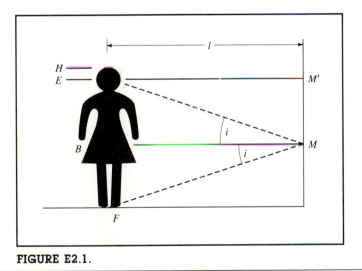

FIGURE E2.1.

What about the light that crosses the surface between two transparent media and is not reflected? What is its behavior? This is described by the *law of refraction*. Note that in reflection only one medium is involved, the ray arises and is reflected back into the same medium, while in refraction two different media are involved. To distinguish the media

TABLE 2.1   Refractive Indices of Various Materials

Gases	
Vacuum	1.0000
Air	1.0003
CO_2	1.0005
Liquid 20°C	
Water	1.333
Ethanol	1.361
CCl_4	1.461
CS_2	1.628
Solids 20°C	
Ice, 0°C	1.310
NaCl	1.54
Glass	1.50–1.80
Crown	1.523
Flint	1.603
Diamond	2.42

one introduces a number characteristic of the medium called the *index of refraction, n.* Later we will see that n is c/v_{med}, where c is the free-space velocity of light and v_{med} is the velocity of the light in the medium. Refractive indices are unique for each particular material and are often used as characterizing parameters for materials. Table 2.1 lists the refractive indices of several common materials.

Generally, as one can see from Table 2.1, gases have refractive indices close to unity, liquids in the range 1.3–1.6, and solids typically greater than 1.4. These generalizations have exceptions, but if one is given a gas with an index of refraction specified as 1.3, for example, one should be wary.

The Law of Refraction.   The angle made by the incident ray, I, that made by the refracted ray, I', and the surface normal at the point of incidence in a refractive process obey the expression

$$n \sin I = n' \sin I' \qquad (2.1)$$

The incident ray, the refracted ray, and the surface normal are all coplanar.

The geometry of refraction is shown in Figure 2.3. The ray incoming from O is refracted at point p in the surface S, and passes through point O'. The angles of incidence and refraction are i and i', respectively. The incoming ray, outgoing ray, and the normal are all in plane P.

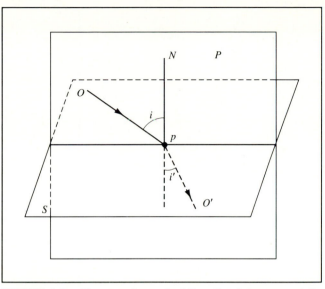

FIGURE 2.3. The ray from O strikes the interface at the surface S between media 1 and 2 at point P and i and i' are fixed by the law of refraction. The normal N and OPO' all lie in plane P.

An example will easily illustrate the application of the law of refraction, which is also known as *Snell's law* in honor of its discoverer.

EXAMPLE 2.2   A light ray strikes the interface between air and water at an angle of 30° with the surface normal. What is the angle made by the refracted ray in the water?

Solution

$$1.000 \times \sin 30° = 1.333 \times \sin I'$$

$$\sin I' = 0.3751$$

$$I' = 22.03°$$

The geometry is shown in Figure E2.2

Generally, the ray will be closer to the normal in the medium with the higher index of refraction n.

If one examines Example 2.2 from the point of view of a ray arising in the water making an angle of 22.03° with the surface normal and passing through B', a point on the ray generated by the ray incident from A' in air, one finds that the ray would travel along precisely the same path as the air-generated ray but in the opposite direction. This is a simple example of a more general optical principle, *optical reversibility*. This principle

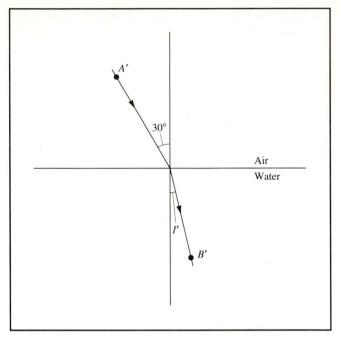

FIGURE E2.2.

states that reversing the direction in which the light is propagated does not alter the path. This principle is used, for example, in some special cases of optical design and in some problems where an end point of a ray and its direction are known and one wants to find the origin of the ray.

If one assumes that the ray in the figure of Example 2.2 arises in the water and makes an angle of 50° with the normal, then

$$\sin I' = \frac{1.333}{1.000} \times \sin 50° = 1.20 \qquad (2.2)$$

Remember now that I' here is the angle made with the surface normal in air. Clearly there is a problem, for the sine of an angle cannot exceed unity. What happens in such a case? As the value of $\sin I'$ approaches unity, the ray in the air comes closer and closer to paralleling the air–water interface after refraction. When $\sin I'$ exceeds unity, the light is constrained to remain in the water, and one has *total internal reflection;* no light escapes from the water in this situation. The light is totally reflected at the interface.

Total internal reflection is often used in optical instrumentation. Consider the prism shown in Figure 2.4. Light enters the prism at the front face parallel to the base and strikes the second face at an angle equal to the angle A of the prism, say, 45°. If $n_{\mathrm{prism}} \sin A > 1.000$, then all the entering light is reflected and passes out through the base of the prism. The second face of the prism acts as a mirror but a mirror that does not need silvering.

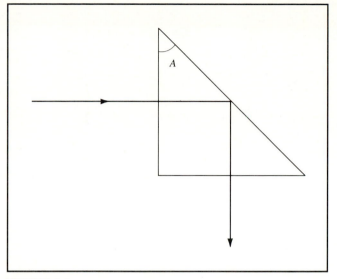

FIGURE 2.4.

Total internal reflection plays a very significant role in optical fibers, as will be seen later, in Chapter 10.

EXAMPLE 2.3

What must be the minimal index of the prism in Figure 2.4 if it is to function as a mirror in the configuration shown?

Solution

The requirement is that

$$n_{prism} \sin 45° > 1.000$$

or

$$n_{prism} > 1.414$$

This condition is met by most solid optical materials.

Spherical Surfaces

Spherical surfaces are of particular importance in optics, as one shall see in what follows, because they permit the formation of images with magnification while typically maintaining high image fidelity. Of equal significance, however, is the fact that spherical surfaces are among the easiest surfaces to produce and test.

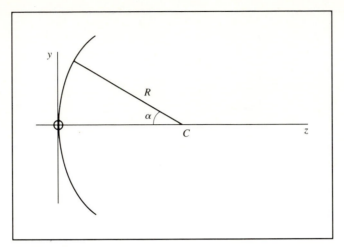

FIGURE 2.5.   The sign convention.

A subject of concern in image analysis is the algebraic sign to be associated with the various quantities that are needed in the analysis of imaging processes. Figure 2.5 illustrates the convention. The origin of the system is always taken at the vertex of the surface being considered, O in the figure. The coordinate system at this point is the usual right-handed Cartesian system with the positive z axis to the right and the positive y axis toward the top of the figure. Angles are also treated traditionally, with a positive angle being one which, when measured from the z axis through the smallest angle, is measured in an anticlockwise fashion. The angle α in the figure is negative.

Spherical Mirrors

Consider now the spherical mirror shown in Figure 2.6. The z axis is the axis of symmetry here and the mirror is rotationally symmetric about this axis, usually called the *optic axis*. A source of light at A is the origin of a ray of light that strikes the mirror at M. The light reflected from the mirror makes equal angles θ with the surface normal. When one views the mirror, one sees the source not at A but at B, and one would like to have the relationship between the position of the source A, the mirror's parameters, and the apparent position of the source B.

The mirror has its center at C and its radius is R. The radius to M defines the surface normal at M and the incident and reflected rays make equal angles, θ, with the surface normal. With θ as the external angle to triangle AMC, one can write

$$\theta = \alpha - \gamma \tag{2.3}$$

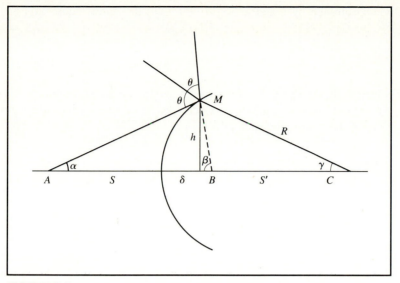

FIGURE 2.6.

since the sign convention makes γ (and β) negative. Using triangle AMB,

$$2\theta = \alpha - \beta \tag{2.4}$$

so that

$$\alpha + \beta = 2\gamma \tag{2.5}$$

and the problem becomes one of defining the angles in terms of the distances along the optic axis.

The height of M above the axis is taken as h, and the axial distance of A from the point of reflection is $s + \delta$, where δ is the small z displacement from the vertex due to the shape of the mirror. If the angles α, β, and γ are small, then these angles can be replaced by their tangents. This small-angle condition is known as the *paraxial approximation*. What is a "small angle"? If the angle is 15° (0.2618 rad), its tangent is 0.2679, and the approximation error is only 2.29%. A second implication of paraxial approximation is that the distance δ in Figure 2.6, the *sagittal depth* of the surface, is also small.

One now replaces α, β, and γ in equation (2.5) with their tangents

$$\alpha = \frac{h}{s + \delta}, \qquad \beta = -\frac{h}{s' + \delta}, \qquad \gamma = \frac{h}{R - \delta}$$

Substituting these in (2.5) yields

$$\frac{h}{s + \delta} + \frac{h}{s' - \delta} = 2\frac{h}{R - \delta} \tag{2.6}$$

Dividing through by h and letting $\delta \rightarrow 0$ gives the mirror equation

$$\frac{1}{s} + \frac{1}{s'} = \frac{2}{R} \tag{2.7}$$

This establishes the position of B, given A and the radius of the mirror R. This equation is valid for both concave and convex surfaces. If one views the mirror now, the source appears to be at B rather than at A.

EXAMPLE 2.4

A point source of light is placed 40 cm from a concave spherical mirror (Figure E2.4). Where does the source appear when the mirror is viewed if the mirror's radius is -10 cm?

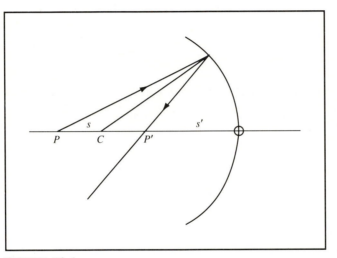

FIGURE E2.4.

Solution

The point source at P is -40 cm from the mirror's vertex, and the radius of curvature of the mirror is -10 cm. Note that the sign is negative since the center of curvature lies to the negative side of the vertex. Using the mirror equation (2.7),

$$-\frac{1}{40} + \frac{1}{s'} = -\frac{2}{10}$$

and $s' = -5.71$ cm.

If one examines Figure E2.4, it can be seen that if P is outside C, the P' must be inside C. Note that points occur pairwise and that, for each *object point P*, there is a corresponding conjugate *image point P'*. As $P \rightarrow -\infty$, P' approaches a value of $R/2$. This

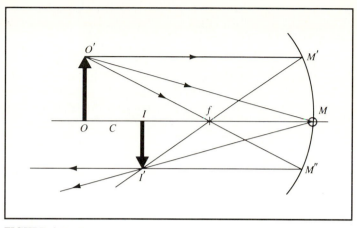

FIGURE 2.7.

limiting value, the image point conjugate with the point at infinity, is known as the *focal length* and the position of this point, $s' = R/2$, is known as the *focal point*. Any ray coming to the mirror from the point at infinity approaches the mirror parallel to the optic axis and, after striking the mirror, goes out along a line passing through the focal point. Similarly, any ray approaching the mirror passing through the focal point before striking the mirror will go out from the mirror parallel to the optic axis as a result of optical reversibility. This gives a method for graphical construction of the image of an extended object as well as simple points on the optic axis.

Consider the concave mirror shown in Figure 2.7. An extended object is represented by the arrow at OO'. Three rays are drawn to establish the point conjugate with the tip of the arrow. The ray parallel to the optic axis $O'M'$, that is, the ray originating at infinity, goes out of the mirror through the focal point f. The ray entering through the focal point $O'M'$ exits the mirror parallel to the optic axis as a consequence of optical reversibility. Finally, the ray that strikes the mirror at its vertex makes an equal angle after reflection. These three rays intersect at I', the point conjugate with O'. Figure 2.8 shows the construction for a convex mirror. The selected rays are precisely those used in Figure 2.7. The difference between the concave and convex mirror cases is that, with the concave mirror, the rays as constructed only apparently converge at I'. Images formed by rays that actually converge are called *real images,* while those images where the convergence is only apparent as in Figure 2.8 are called *virtual images*. With a real image, a screen placed at the image position will have the image projected on it, while with a virtual image, one will not be able to position a screen in such a way as to have the image projected on it.

The characterization of the image involves specifying, in addition to the position of the image, whether the image is larger or small, upright or inverted, and real or virtual. For example, the image in Figure 2.7 is smaller, inverted, and real, while that in Figure 2.8 is smaller, erect, and virtual.

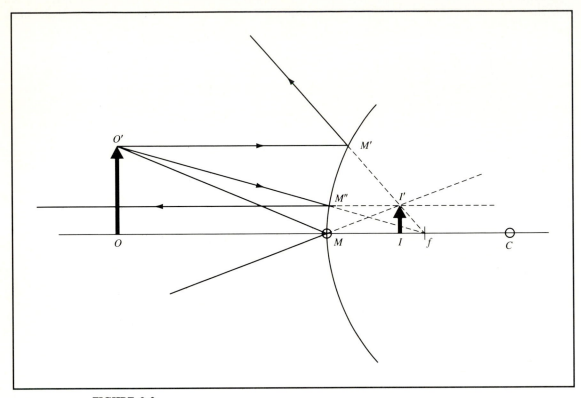

FIGURE 2.8.

The ratio of the size of the image to the size of the object is the magnification of the system. Inspection of Figures 2.7 and 2.8 using the similarity of the triangles $OO'M$ and $II'M$ shows that the magnification is given by

$$m = -\frac{s'}{s} \tag{2.8}$$

The negative sign is important in that a negative magnification implies an inverted image. The size of the image is the product of the size of the object and the magnification.

EXAMPLE
2.5

A candle 5 cm tall stands 30 cm in front of a convex spherical mirror of radius 12 cm (Figure E2.5). Locate and characterize the image.

Solution Using the mirror equation (2.7),

$$\frac{1}{-30} + \frac{1}{s'} = \frac{2}{R} = \frac{1}{f} = \frac{1}{6}$$

$$s' = +5 \text{ cm}$$

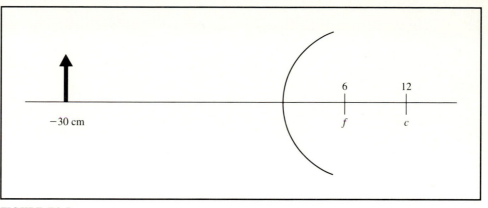

FIGURE E2.5.

and the magnification is

$$m = -\frac{5}{-30} = \frac{1}{6}$$

and the image of the candle is 5/6 cm tall, erect, and virtual.

Spherical Refracting Surfaces

The relationships between the quantities involved in refraction at a spherical surface can be found easily by methods analogous to those for spherical reflecting surfaces. Figure 2.9 illustrates the geometry. A ray arises at a point source A on the optic axis and strikes the spherical interface separating the medium characterized by index n_1 from that characterized by n_2 at point F. The ray is refracted and intersects the optic axis at B. Again one seeks the relationship among the distance s' from B to the vertex of the spherical surface, the source distance s, and the radius of the surface R.

The procedure here is similar to that used with spherical reflectors. One uses the external angles θ and γ here, where

$$\theta = \alpha - \gamma$$
$$\gamma = \beta - \theta' \tag{2.9}$$

and

$$\theta = \beta - \gamma \tag{2.10}$$

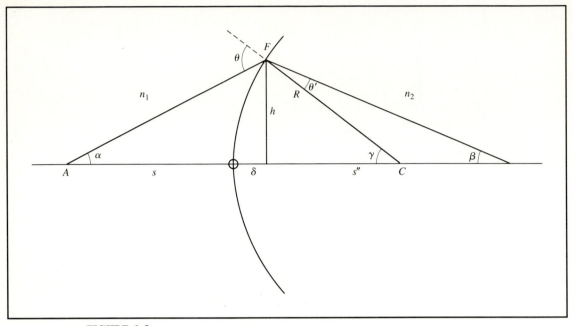

FIGURE 2.9.

Here one uses Snell's law

$$n_1 \sin \theta = n_2 \sin \theta'$$

since again the surface radius vector establishes the surface normal. In the paraxial approximation,

$$n_1 \theta = n_2 \theta' \tag{2.11}$$

and substituting from (2.9) and (2.10),

$$n_1(\alpha - \gamma) = n_2(\beta - \gamma) \tag{2.12}$$

which can be rearranged

$$n_1\alpha - n_2\beta = -(n_2 - n_1)\gamma \tag{2.13}$$

Again, applying the paraxial approximation, one replaces the angles with their tangents,

$$\frac{n_1 h}{s + \delta} - \frac{n_2 h}{s' - \delta} = -(n_2 - n_1)\frac{h}{R - \delta}$$

As $\delta \rightarrow 0$ one gets

$$-\frac{n_1}{s} + \frac{n_2}{s'} = \frac{(n_2 - n_1)}{R} \qquad (2.14)$$

which is the equation governing refraction at a single spherical interface. As with reflection, the relationship between s and s' is independent of the height at which the ray strikes the surface within the constraints of the paraxial approximation.

Again one has pairs of object–image conjugates. One significant difference here is that there are two different focal points corresponding to the points at infinity at either side of the refracting surface. The first focal point or object focal point is the point conjugate with an object at infinity as shown in Figure 2.10a. The second focal point or image focal point is the point conjugate with the image point at infinity as shown in Figure 2.10b. Applying equation (2.14) with $s \rightarrow -\infty$, one gets

$$-0 + \frac{n_2}{s'} = \frac{(n_2 - n_1)}{R} = \frac{n_2}{f_1} \qquad (2.15)$$

and

$$\frac{n_2}{f_1} = \frac{(n_2 - n_1)}{R} \qquad (2.16)$$

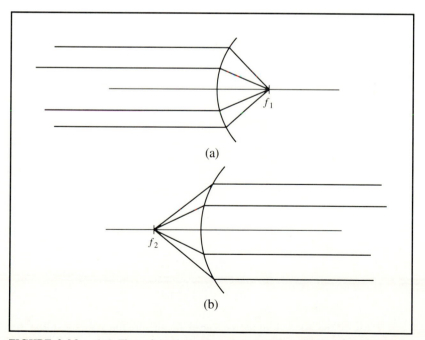

(a)

(b)

FIGURE 2.10.   (a) The object focal point; (b) the image focal point.

Similarly,

$$\frac{n_1}{f_2} = -\frac{(n_2 - n_1)}{R} \tag{2.17}$$

These equations can serve as definitions of the focal lengths of the surface. There is a second way of specifying the refracting properties of a spherical surface, and that is by specifying the surface power. The *surface power* is the reciprocal of the focal length. When the focal length is measured in meters, the power is in *diopters,* usually denoted D. A surface with a 10-cm focal length (0.1 m) has a power of 10 D while a 5-cm focal length surface has a power of 20 D. The larger the power, the stronger the focal properties of the surface.

EXAMPLE 2.6

An important example of a single spherical refracting surface is the cornea of the eye. The cornea, with a radius of curvature of 7.5 mm separates the outside (air) from the ocular medium with an index of 1.336. While the cornea is only the first refracting element of the eye, the internal crystalline lens being the second, the cornea is the strongest. What is its focal length and power if the cornea's index of refraction is 1.376?

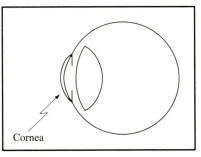

FIGURE E2.6.

Solution

Using equation (2.16) one finds

$$\frac{1}{f_1} = \left[\frac{(n_2 - n_1)}{n_2}\right]\left(\frac{1}{R}\right)$$

$$= \left(\frac{1.376 - 1.0}{1.376}\right)\left(\frac{1}{7.5}\right)$$

$$f_1 = 27.45 \text{ mm}$$

This is the focal length measured in air. In the eye itself the focal length is a "reduced" length f/n_1. In sketching systems such as the eye, the reduced focal length must be used. The question of reduced distances will be treated in more detail when paraxial ray tracing is addressed in Chapter 3.

To get the surface power, the inverse of the reduced focal length, f_1, must be expressed in meters, and

$$P = \frac{1.336}{0.02745} \text{ m} = 49.8 \text{ D}$$

Thin Lenses

Rarely does one deal with a single refracting surface. Spherical surfaces are usually combined to form *lenses*. Here we will treat a simple thin lens while thick lenses and combinations of thin lenses will be treated within the framework of paraxial ray tracing. A thin lens is formed by two refracting surfaces a small distance δ apart, separating the lens medium from the surrounding material, most commonly air. The image formed by the first surface becomes the object for the second surface, and by a "relay" procedure, one gets the lens equation.

Figure 2.11 shows a thin lens. The image formed by the first surface becomes the object for the second surface. The radii of the surfaces are R_1 and R_2, respectively, the index of the lens is n, and the index of the medium surrounding the lens is n_{med}. Let the position of the image formed by surface R_1 be at s'' as in the figure. Using equation (2.14),

$$-\frac{n_\text{m}}{s_\text{o}} + \frac{n_\text{m}}{s''} = \frac{n_\text{g} - n_\text{m}}{R_1} \tag{2.18}$$

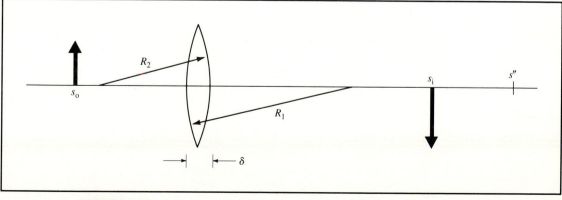

FIGURE 2.11.

The image at s'' will now become the object for the refraction at the second surface and the conjugates are "relayed" to the second surface:

$$-\frac{n_g}{s''} + \frac{n_m}{s_i} = \frac{n_m - n_g}{R_2} \tag{2.19}$$

where, as a consequence of the thin-lens approximation, the central thickness $\delta \rightarrow 0$. On eliminating the intermediate image position s'' from (2.18) and (2.19), one gets

$$-\frac{n_m}{s_o} + \frac{n_m}{s_i} = (n_g - n_m)\left(\frac{1}{R_1} - \frac{1}{R_2}\right) \tag{2.20}$$

which is the *thin-lens equation*. Again, this equation gives the relationship between the position of an object s_o and that of its image s_i in terms of the optical structure acting on the light from the object.

As with a single refracting surface, two focal lengths exist in the system. The focal length, f_o, is conjugate with an object at infinity,

$$\frac{n_m}{f_o} = (n_g - n_m)\left(\frac{1}{R_1} - \frac{1}{R_2}\right) \tag{2.21}$$

while the image focal length, f_i, is the position conjugate with an image at infinity and for a thin lens f_i is simply $-f_o$.

Typically n_m is unity since lenses most commonly act in air. Equation (2.21) with $n_m = 1$,

$$\frac{1}{f_o} = \phi = (n_g - 1)\left(\frac{1}{R_1} - \frac{1}{R_2}\right) \tag{2.22}$$

gives the lens power, ϕ, in diopters if the lengths are in meters. This equation is known as the *lensmaker's equation*.

The lens equation may then also be written in the form

$$-\frac{n_m}{s_o} + \frac{n_m}{s_i} = \phi = \frac{1}{f_o} \tag{2.23}$$

EXAMPLE 2.7   A thin lens of crown glass, $n = 1.52$, is to have a power of 10 D in air. If R_1 is 0.02 m, what is R_2? What is the focal length?

Solution   Using equation (2.21),

$$10 = (1.52 - 1.00)\left(\frac{1}{0.02} + \frac{1}{R_2}\right)$$

$$R_2 = -0.0325 \text{ m}$$

and using equation (2.22),

$$f = \frac{1}{\phi} = \frac{1}{10} = 0.1 \text{ m}$$

Example 2.7 illustrates an important point. For a given power there is essentially an infinity of choices of the pairs R_1, R_2 that will produce the required power. The image fidelity is dependent on the choice of the appropriate pair of values, and this problem will be considered later, in Chapter 4.

EXAMPLE 2.8   A -10-D lens in air forms an image of an object placed 25 cm in front of it. Locate the image.

Solution   Using equation (2.23),

$$-\frac{1}{-0.25 \text{ m}} + \frac{1}{s_i} = -10 \text{ D}$$

Note that since the power was expressed in diopters, it is necessary to use -0.25 m for the object distance to maintain consistency in units since 1 diopter $= 1 \text{ m}^{-1}$:

$$s_i = -0.0714 \text{ m} \simeq -7.14 \text{ cm}$$

The negative sign shows that the image is also in front of (to the left of) the lens.

EXAMPLE 2.9   A lens is known to have a focal length of 30 cm in air. An object is placed 50 cm to the left of the lens. Locate the image.

Solution   Using the thin-lens equation (2.23),

$$-\frac{1}{-50} + \frac{1}{s_i} = \frac{1}{30}$$

$$s_i = 75.00 \text{ cm}$$

that is, 75.00 cm to the right of the lens.

One can use graphical construction with lenses to verify the location of the image. The procedure is similar to that used with mirrors or a single refracting surface. Figure 2.12a shows the object–image conjugates from Example 2.9. The ray from the tip of the

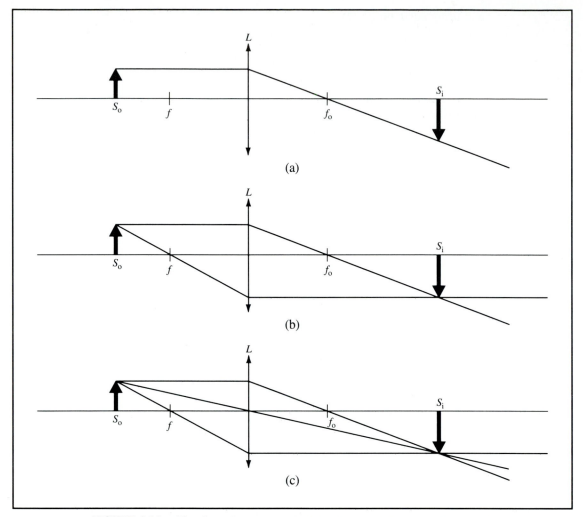

FIGURE 2.12.   The lens is represented by the double-ended arrow L.

object parallel to the optic axis, that is, a ray from infinity, passes out of the system through the focal point f_o after crossing the lens as shown in the figure. In Figure 2.12b the ray passing from the tip of the object through the object-side focal point is added, and this is conjugate with the point at infinity so that, after crossing the lens, it exits the system parallel to the optic axis. Finally, the ray crossing the lens at the optic axis is undeviated, as shown in Figure 2.12c. The image can be seen to be inverted, larger, and real.

The expression for the magnification of the system can be gotten from Figure 2.12c. The magnification, μ, is given by h'/h. Similar triangles yields

$$\mu = \frac{s}{s'} = \frac{h'}{h} \qquad\qquad (2.24)$$

EXAMPLE   What is the magnification of the image in the preceding example?
2.10

Solution   The object distance is $s_o = -50$ cm, and the image distance was found to be $s_i = 75$ cm. Thus

$$\mu = \frac{75}{-50} = -1.50$$

The image is larger, and the negative sign shows that it is inverted.

Newton's Lens Equation

Figure 2.13 is identical with Figure 2.12b, where the object–image pair is located by rays through the two focal points. Note in the figure that $OL = h$ and that $OL' = h'$. Using similar triangles one can write

$$\frac{h}{x} = \frac{h'}{f}$$

and that

$$\frac{h'}{x'} = \frac{h}{f_o}$$

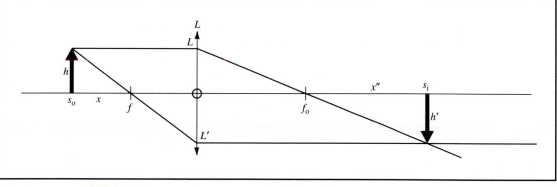

FIGURE 2.13.

Combining these yields

$$\frac{h}{h'} = \frac{x}{f} = \frac{f_o}{x'}$$

or

$$xx' = ff_o \qquad (2.25)$$

which is Newton's form of the lens equation. This form of the lens equation is applicable to both thin lenses as here and to thick lenses, although with thick lenses the position from which the focal length is measured is typically not the center of the lens. Note that since $f = -f_o$, Newton's lens equation can be written

$$xx' = -f_o{}^2 \qquad (2.26)$$

EXAMPLE 2.11

Solve Example 2.9 using Newton's form of the lens equation.

Solution

The object is 50 cm to the left of the lens and the focal length is 30 cm, which means that x is given by

$$x = -50 - (-30) = -20 \text{ cm}$$

and using equation (2.26) gives

$$(x')(-20) = -(30)^2$$

or

$$x = 45 \text{ cm}$$

The total distance from the lens to the image is

$$s_o = f_o + x' = 30 \text{ cm} + 45 \text{ cm} = 75 \text{ cm}$$

in complete agreement with what was found in Example 2.9. Note that since the magnification is given by h'/h, the magnification can be written in terms of the Newtonian variables as

$$\mu = -\frac{h'}{h} = -\frac{x'}{f_o} = -\frac{x}{f} \qquad (2.27)$$

so that the magnification in the example is

$$\mu = -\frac{45}{30} = -\frac{-30}{-20} = -1.5$$

as was found in Example 2.10.

PROBLEMS

2.1. In ancient times the rectilinear propagation of light was used to measure the height of objects by comparing the length of their shadows with the length of the shadow of an object of known length. A staff 2 m long when held erect casts a shadow 3.4 m long, while a building's shadow is 170 m long. How tall is the building?

2.2. An intense point source of light illuminates a plane white wall 10 m away. An opaque disk is placed between the source and the wall. What is the relationship between the area of the disk and the area of the shadow as a function of the disk's position?

2.3. The source in the preceding problem is allowed to expand. Describe the shadow of the disk as the source increases in size.

2.4. An artificial satellite is positioned above the equator in fixed geocentric orbit (it remains fixed in position over a point on the earth). For how long after sunset will it remain visible? Assume the earth's radius is 4000 mi, the satellite's altitude is 22,000 mi, and the observer is at the equator.

2.5. Light from a water medium with $n = 1.33$ is incident upon a water–glass interface at an angle of 45°. The glass index is 1.50. What angle does the light make with the normal in the glass?

2.6. Diamonds sparkle because they internally reflect light incident on them. The top surface of a diamond is a plane, and it is cut to have a point at the bottom with included angle θ. What is θ so that light normally incident on the upper surface is totally reflected internally?

2.7. Two plane mirrors are set parallel to each other at a distance s. An object is inserted between the mirrors at a distance s' from one of them. Where do the images appear?

2.8. An object is s cm in front of a plane mirror. Describe the locus of the image as the mirror is rotated about an axis normal to the perpendicular from the object at its intersection with the mirror.

2.9. Some shaving mirrors and makeup mirrors are made concave to give a magnified image. How large should the radius of curvature be if the distance to the face is s cm? Give a reasonable range for your answer.

2.10. Some automotive rear view mirrors are convex spherical mirrors. These often have "Objects in mirror are closer than they appear" engraved on them as a notice. If the radius of curvature is 60 m and a car is 30 m behind the mirror, where does it appear to be when viewed in the mirror?

2.11. If a car approaches the mirror in the problem above at a relative speed of 2.7 m/s (about 6 mph), how fast does it appear to be approaching in the mirror?

2.12. Dentists use concave mirrors to view teeth when filling them. The radius of curvature of such a mirror is 5 cm and the mirror is held 1 cm from the tooth being filled. How much larger does the tooth appear?

2.13. An object is 100 cm in front of the cornea described in Example 2.6. Where is the image of this object formed?

2.14. A goldfish swims 10 cm from the side of a spherical bowl of water of radius 20 cm. Where does the fish appear to be? Does it appear larger or smaller?

2.15. What are the radii of curvature of a 15-D flint glass ($n = 1.605$) equiconvex lens? Of a -15-D equiconcave lens? If the lens is 3 cm in diameter, will an equiconvex 15-D lens or a planoconvex 15-D lens be thinner?

2.16. An object 1 cm high is 30 cm in front of a thin lens with a focal length of 10 cm. Where is the image and what is its magnification? Verify your result by graphical construction of the image.

2.17. Repeat Problem 2.16 using Newton's form of the lens equation.

2.18. A -5-D lens has an object 1 cm tall 10 cm before it. Locate the image and specify the magnification. Verify your answer by graphical construction.

CHAPTER 3

Paraxial Ray Tracing

You have already seen that the lens equation in one or another of its forms can be used to find the position of an image in an optical system. This is usually easy for a thin lens, for example, where a single object point and the value of the lens power or focal length can be used to find the conjugate image point. With thick lenses or with a system of lenses, this procedure can still be used in an iterated surface-by-surface fashion to locate object–image conjugates; the process will become tedious but it will work. This approach has two failings, however; the need for iterated calculations in any but the simplest cases and the lack of information about how far any ray is from the optic axis at any arbitrary point in the system. The latter failing prevents one from having some sense about the physical dimensions required of the lenses in a system of lenses.

A second problem is not so evident, but it exists as well. Where is your reference point? It does in fact move from surface to surface with the iteration, and in general, one has some difficulty relating changes in the object point to changes in the image point without carrying along a considerable burden in terms of dummy intermediate variables.

Remember that here we are working with a *paraxial* model. The results will only approximate those of an exact calculation within the constraints pointed out earlier in the derivation of the lens equation. Still, these results are for the most part more than accurate enough to describe the properties of a lens system. The calculations are more accurate in a full trigonometric trace, but the issue of a full trace is left for the discussion of optical design.

The process of locating object–image conjugates is one of relating input variables to output variables. In this case that constitutes a mapping of points in object space with its origin at the first vertex of the system onto points in image space with its origin at the last vertex of the system. Each of these two spaces fills all space since virtual objects and images may lie on what one may consider the ''wrong side of the system.''

The System Matrix

The elements of an optical system are characterized by their surface radii and their thicknesses. These values, along with the refractive indices of the elements making up the system, are what give each system its unique properties. Table 3.1 is a table of these values for a simple model lens system. Figure 3.1 is a drawing of the lens. One finds specified in this table everything needed at this stage to characterize the lens system. Tables such as this are one common way that lens data are specified. What one would like is a general procedure that will give us the same information as provided by Newton's lens equation or the thin-lens equation for any lens system. In addition, one would like to have information about the ray heights at any point in the system and be able to relate these to a known and fixed coordinate system.

To begin, look at Table 3.1. A ray entering the lens system at some height y has positive value if it is above the optic axis (z axis). Surfaces have the effect of changing the direction of the ray but do not alter its height at the surface. That is, the ray is continuous in the mathematical sense with a discontinuity in its first derivative at a surface, in other words, it abruptly changes its slope. In contrast, the regions between successive surfaces introduce no discontinuities into the ray or its first derivatives, but will involve a change in

TABLE 3.1 A Simple Air-Spaced Doublet

Surface	1		2		3		4
Radius	12.00		−12.0		−20.0		20.0
Thickness		0.75		1.00		0.48	
Index		1.50		1.00		1.60	

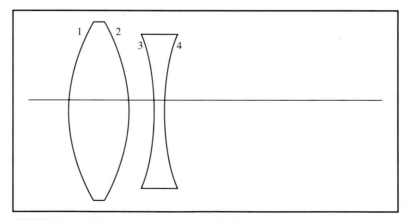

FIGURE 3.1. The lens specified in Table 3.1.

the y value of the ray between surfaces. We will consider first the changes in passing between surfaces and then the effect of the surfaces on a ray, finally combining the two.

The sign convention remains that established in Chapter 2. The origin is taken as the pole of the next surface, that is, at the point where the optic axis intersects the surface. The y coordinate is positive above the axis, the z coordinate is positive to the right, and the x coordinate is positive below the plane of the diagram. The coordinate system is a right-hand cartesian system. All angles are measured with respect to the optic axis so that angles measured in an anticlockwise sense from the optic axis are positive.

What will happen to a ray as it passes from surface i to surface $i + 1$ in the system? Assume that the ray is at height y_i at surface i. If the ray makes an angle α_i with the optic axis, then the height at surface $i + 1$ is given by

$$y_{i+1} = y_i + t \tan \alpha_i \tag{3.1}$$

and in the paraxial approximation $\tan \alpha \approx \alpha$ so that

$$y_{i+1} = y_i + t\alpha_i \tag{3.2}$$

At the second surface α is unchanged and

$$\alpha_{i+1} = \alpha_i \tag{3.3}$$

What happens to α as the ray crosses the surface? The y value remains unchanged since there is no height change for a ray in crossing the surface; however, α does change because of the refraction. The geometry of this change is shown in Figure 3.2. In the

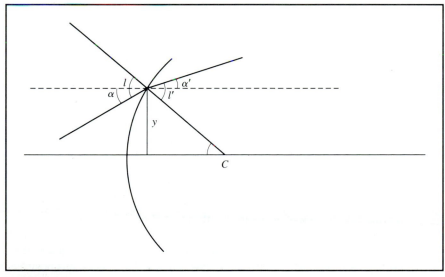

FIGURE 3.2.

figure C is the center of curvature of the refracting surface. Note that I and I' are the angles of incidence and refraction, but that it is the *angle made with the optic axis* that is the variable of interest, just as in (3.1).

Snell's law gives

$$n \sin I = n' \sin I' \tag{3.4}$$

where n is the index to the left of the surface and n' that to the right of the surface. In the paraxial approximation this translates to

$$nI = n'I' \tag{3.5}$$

Note that in the figure the dashed line zz' is parallel to the optic axis so that

$$I = \alpha + \theta \tag{3.6}$$

and

$$I' = \alpha' + \theta \tag{3.7}$$

and in the paraxial approximation

$$\theta = \frac{y}{R} \tag{3.8}$$

Combining (3.5)–(3.8) one gets

$$n(\alpha + \theta) = n\left(\alpha + \frac{y}{R}\right) = n'(\alpha' + \theta) = n'\left(\alpha' + \frac{y}{R}\right) \tag{3.9}$$

and

$$n'\alpha' = n\alpha - y\left(\frac{n' - n}{R}\right) \tag{3.10}$$

This, combined with the fact that y is fixed at a surface and does not change when only the direction of the ray changes,

$$y = y' \tag{3.11}$$

summarizes the effect of a surface of radius R on an incident ray.

It is important to bear in mind the sign convention for surfaces here. If the center of curvature is to the right in the conventional left-to-right presentation of the optical system, it is positive. Another way of stating this is to say that a surface is positive if the center of curvature is on the same side as the outgoing ray.

There are two equations governing rays in each case. The translation of rays between surfaces is summarized by the pair of equations (3.2) and (3.3) while equations (3.10) and (3.11) summarize the effect of crossing an interface. The variables are the same in each case, the displacement of the ray with respect to the optic axis y and the angle made by the ray with respect to the optic axis α. Since the variable α appears in (3.9) with the index of refraction of the medium in which α is being measured as a multiplier, it is more common to use $\boldsymbol{\alpha}_i = n_i \alpha_i$ as the variable. Equation (3.2) now reads

$$y_{i+1} = y_i + \frac{t}{n_i} \boldsymbol{\alpha}_i \tag{3.12}$$

The thickness variable t can similarly be replaced by the "reduced thickness," the physical thickness weighted by the reciprocal of the index of refraction of the medium, $\mathbf{t} = t/n_i$, so that the equation will read

$$y_{i+1} = y_i + \mathbf{t}\boldsymbol{\alpha}_i \tag{3.13}$$

Pairs of equations such as (3.9) and (3.10), and (3.3) and (3.4) can be efficiently represented in matrix notation using 2×2 matrices and column vectors to present the variables. In this notation the translation process of (3.3) and (3.13) can be written

$$\begin{bmatrix} \boldsymbol{\alpha}_{i+1} \\ y_{i+1} \end{bmatrix} = \begin{bmatrix} n_i \alpha_{i+1} \\ y_{i+1} \end{bmatrix} = \begin{bmatrix} 1 & 0 \\ \mathbf{t}_i & 1 \end{bmatrix} \begin{bmatrix} \boldsymbol{\alpha}_i \\ y_i \end{bmatrix} \tag{3.14}$$

where n_i is the index of the medium *following* the ith surface and t_i is the thickness of the medium following the ith surface.

In a similar fashion, the pair of equations (3.10) and (3.11) can be represented

$$\begin{bmatrix} \boldsymbol{\alpha}_i' \\ y_i \end{bmatrix} = \begin{bmatrix} 1 & -\dfrac{n'-n}{R} \\ 0 & 1 \end{bmatrix} \begin{bmatrix} \boldsymbol{\alpha}_i \\ y_i \end{bmatrix} \tag{3.15}$$

where $\boldsymbol{\alpha}_i'$ is the angle of the ray after refraction at the ith surface, again weighted by the index of the medium.

It is now easy to see how one might trace a ray through a lens system. If the angle $\boldsymbol{\alpha}_o$ and the ray height at the initial surface of a system are known, then one can get the angle and height of the ray at any other position in the system by the repeated application of the *refraction matrix*

$$[\mathscr{R}]_j = \begin{bmatrix} 1 & -\dfrac{n'-n}{R_j} \\ 0 & 1 \end{bmatrix} \tag{3.16}$$

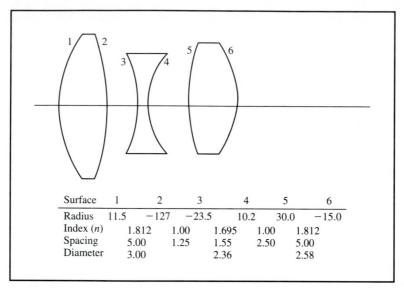

Surface	1		2		3		4		5		6
Radius	11.5		−127		−23.5		10.2		30.0		−15.0
Index (n)		1.812		1.00		1.695		1.00		1.812	
Spacing		5.00		1.25		1.55		2.50		5.00	
Diameter		3.00				2.36				2.58	

FIGURE 3.3.

where $[\mathscr{R}]_j$ is the matrix for the jth surface with radius R_j and the *translation* matrix

$$[\mathscr{T}]_k = \begin{bmatrix} 1 & 0 \\ \mathbf{t}_k & 1 \end{bmatrix} \qquad (3.17)$$

where $\mathbf{t}_k$ is the reduced spacing following the kth surface.

Assume now that one is evaluating a lens such as that shown in Figure 3.3. The lens has three elements with six surfaces and five translational spacings. The surfaces are numbered 1 to 6 in order. The overall ray path is given by

$$\begin{bmatrix} \boldsymbol{\alpha}_f \\ y_f \end{bmatrix} = [\mathscr{R}]_6[\mathscr{T}]_5 \cdots [\mathscr{T}]_2[\mathscr{R}]_2[\mathscr{T}]_1[\mathscr{R}]_1 \begin{bmatrix} \boldsymbol{\alpha}_i \\ y_i \end{bmatrix} \qquad (3.18)$$

where $\boldsymbol{\alpha}_i$ and y_f are the weighted angle and the height of the ray emerging from the system. Of course, one may interrupt this process at any point within the system to find the ray angle and height if the intermediate height or angle must be found, as in placing an aperture stop, for example.

The matrix product given by

$$[\mathbf{S}] = [\mathscr{R}]_6[\mathscr{T}]_5 \cdots [\mathscr{T}]_1[\mathscr{R}]_1 \qquad (3.19)$$

where [S] is the final 2×2 product matrix called the *system matrix*. The system matrix is represented as

$$[\mathbf{S}] = \begin{bmatrix} b & -a \\ -d & c \end{bmatrix} \tag{3.20}$$

where the negative signs are introduced at this point to simplify the final equations.

The important point here is that *all the paraxial properties of the system, however complex the system may be, are bound up in the four elements of the system matrix.* This is a significant result in that no further reference need be made to the lens system once the system matrix has been found.

There is a convenient check of the system matrix in that it is the product of only refraction and translation matrices. Each of these has a determinant of 1. That is,

$$|\mathcal{R}| = 1 \tag{3.21}$$

and

$$|\mathcal{T}| = 1 \tag{3.22}$$

By use of the theorem from linear algebra that states that the product of a series of square matrices has a determinant equal to the product of the determinants of the component matrices, one can immediately see that *the determinant of the system matrix is* 1.

EXAMPLE 3.1

The Cooke triplet in Figure 3.3 is made up of three elements. Find the system matrix for the first element.

Solution

The system matrix for the first lens has three components

$$[\mathbf{S}] = [\mathcal{R}]_2 [\mathcal{T}]_1 [\mathcal{R}]_1$$

which are given by

$$[\mathcal{R}]_1 = \begin{bmatrix} 1 & -\dfrac{1.812 - 1.000}{11.5} \\ 0 & 1 \end{bmatrix} = \begin{bmatrix} 1 & -0.0706 \\ 0 & 1 \end{bmatrix}$$

$$[\mathcal{T}]_1 = \begin{bmatrix} 1 & 0 \\ \dfrac{5.00}{1.812} & 1 \end{bmatrix} = \begin{bmatrix} 1 & 0 \\ 2.7594 & 1 \end{bmatrix}$$

$$[\mathcal{R}]_2 = \begin{bmatrix} 1 & -\dfrac{1.000 - 1.812}{-127} \\ 0 & 1 \end{bmatrix} = \begin{bmatrix} 1 & -0.0064 \\ 0 & 1 \end{bmatrix}$$

Note the signs carefully and be sure you know how they are gotten. The most common errors encountered in such calculations are those associated with the signs of the matrix elements or poor arithmetic.

[S] is then given by

$$[\mathbf{S}] = \begin{bmatrix} 0.9824 & -0.0758 \\ 2.7594 & 0.8052 \end{bmatrix}$$

As a check, $|\mathbf{S}| = 1$.

It is important to note the ordering of the elements of the system matrix. The first element to act on the incoming ray is the first surface of the system. Note in Example 3.1 that the ordering of the elements of the system matrix provides for the ordered action of the elements of the lens on the incoming ray. Check the order of the matrices against the successive actions of the elements of the system on the ray. The first surface gives the right-hand matrix; the second appears to its left, and so on.

Imaging

Since all the paraxial properties of the lens system are included in the system matrix, the imaging properties must derive from the elements of the system matrix. To find the imaging relationships we must introduce dummy surfaces. These dummy surfaces are plane surfaces with no index change across them. In this case the dummy surfaces will be taken at the object position and the image position as shown in Figure 3.4. The "thickness" of the first element is the distance between the dummy plane and the first pole. There is no restriction on the placement of dummy surfaces, and such surfaces are often inserted into the lens itself in order to locate particular apertures. Equation (3.18) becomes in this case

$$\begin{bmatrix} \boldsymbol{\alpha}_f \\ y_f \end{bmatrix} = \begin{bmatrix} 1 & 0 \\ \mathbf{s} & 1 \end{bmatrix} \begin{bmatrix} b & -a \\ -d & c \end{bmatrix} \begin{bmatrix} 1 & 0 \\ -\mathbf{s}_o & 1 \end{bmatrix} = \begin{bmatrix} \boldsymbol{\alpha}_o \\ y_o \end{bmatrix} \tag{3.23}$$

Note that the sign of $\mathbf{s}_o$ is negative because we take the origin of the object space at the first pole of the system, the first pole being the intersection of the first surface of the system with the optic axis. Similarly, $\mathbf{s}$ is taken with respect to the last pole of the system, and this then becomes the origin of image space. Remember that $\mathbf{s}_o$ and $\mathbf{s}$ are weighted distances and are equal to s_o/n_o and s/n, respectively.

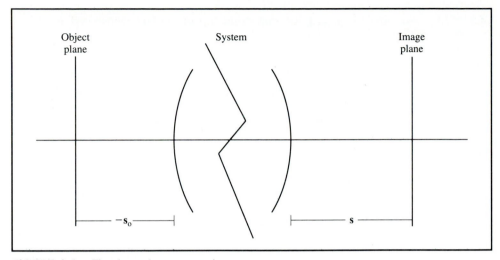

FIGURE 3.4. The imaging geometry.

Equation (3.23) reduces to

$$\begin{bmatrix} \alpha_f \\ y_f \end{bmatrix} = \begin{bmatrix} b + as_o & -a \\ -d - cs_o + s(b + as_o) & c - as \end{bmatrix} \begin{bmatrix} \alpha_o \\ y_o \end{bmatrix} \qquad (3.24)$$

If the dummy surfaces are object–image conjugates as has been assumed, then the equation relating y_f to y_o, that is,

$$y_f = \alpha_o[-d - cs_o + (b + as_o)] + y_o(c - as) \qquad (3.25)$$

must be independent of the angle at which the ray enters the system. That is, if imaging is to take place, then all rays leaving a single point on the object at y_o and passing through the system must also pass through y_f regardless of the angle at which they leave the object point as long as the angles are within the paraxial ray angle assumption. So that there is no dependence on α_o, the coefficient of α_o in equation (3.25) must be zero:

$$-d - cs_o + s(b + as_o) = 0 \qquad (3.26)$$

or

$$s = \frac{d + cs_o}{b + as_o} \qquad (3.27)$$

which is the lens equation in this approach. One should remember that the two origins are at the poles of the surfaces and that s and s_o are reduced optical distances.

EXAMPLE
3.2

An object is placed 200 mm to the left of the lens whose system matrix was found in Example 3.1. Locate the image.

Solution

The matrix elements are

$$a = 0.0758$$
$$b = 0.9824$$
$$c = 0.8052$$
$$d = -2.7594$$

Notice the signs of a and d and remember how the system matrix was defined in equation (3.20). Now $\mathbf{s}_o = -200/1.000$, since the object is to the left of the first pole, and one then gets

$$\mathbf{s} = \frac{s}{1.000} = \frac{-2.7594 + (0.8052)(-200)}{0.9824 + (0.0758)(-200)} = 11.5534 \text{ mm}$$

The image is then 11.5534 mm to the right (positive side) of the second (last) pole of the system.

One should ask what other information can be gotten from equation (3.25). Since the coefficient of $\boldsymbol{\alpha}_o$ is zero, equation (3.25) reduces to

$$y_f = (c - a\mathbf{s})y_o \tag{3.28}$$

The quantity that relates the size of the image y_f to that of the object y_o is the *magnification* μ. One gets

$$\mu = (c - a\mathbf{s}) \tag{3.29}$$

and when one knows the system matrix and the image position, one knows the magnification and therefore the size of the image. Using equations (3.26) and (3.29), the matrix in equation (3.24) now takes the form

$$\begin{bmatrix} b + a\mathbf{s}_o & -a \\ 0 & \mu \end{bmatrix} \tag{3.30}$$

The condition of unit determinant applies to this matrix since it is the product of matrices with unit determinants, and this gives the condition

$$(b + a\mathbf{s}_o)\mu = 1 \tag{3.31}$$

Thus in this approach the magnification can be found in terms of either the object distance or the image distance, and, given the system matrix, it is not necessary to find the image distance in order to know the magnification for a given object and

$$\mu = \frac{1}{b + a s_o} \tag{3.32}$$

This is a very useful expression since by comparing equation (3.32) with equation (3.27) one can see that the magnification is found as part of the process for finding the image. Thus

$$s = \mu(d + c s_o) \tag{3.33}$$

becomes an alternative form for the imaging equation.

EXAMPLE 3.3   Use equations (3.29) and (3.32) to find the magnification for Example 3.2 and verify that the result is consistent.

Solution   From equation (3.29)

$$\mu = c - a\mathbf{s} = 0.8052 - 0.0758 \times \frac{11.5534}{1.000}$$

$$= -0.0705$$

From equation (3.32)

$$\mu = \frac{1}{b + a s_o} = \frac{1}{0.9825 + 0.0758(-200/1.000)}$$

$$= -0.0705$$

and the results are consistent. The negative sign for the magnification simply means that the image is inverted, as one would expect.

Cardinal Points

There is a set of six points that, as we will see, characterizes an optical system in a most useful way. They are the principal or unit planes, the nodal points, and the foci. Knowing these six points allows one, for example, to construct object–image pairs graphically without the complexity of drawing the full system.

The *principal or unit planes* are conjugate planes within a lens system between which the magnification is unity. Since the magnification is given by

$$\mu = c - a\mathbf{s} = \frac{1}{b + a\mathbf{s_o}}$$

setting $\mu = 1$ yields

$$\mathbf{u} = \frac{c - 1}{a} \tag{3.34}$$

which is the image principal plane, and

$$\mathbf{u_o} = \frac{1 - b}{a} \tag{3.35}$$

which is the object principal plane.

The *nodal points* are points on the optic axis taken so that an incoming ray appearing to make an angle θ with the optic axis at the object nodal point will appear in image space to arise from the image nodal point making the same angle with the optic axis. Figure 3.5 illustrates this. The nodal points are gotten from equation (3.24) under the condition that $\alpha_f = \alpha_o$ and $y_f = y_o = 0$. The nodal points are points, in contrast to the unit planes, which are truly planes. If equation (3.24) is written in the form

$$\begin{bmatrix} n\alpha_f \\ y_f \end{bmatrix} = \begin{bmatrix} \dfrac{1}{\mu} & -\phi_s \\ 0 & \mu \end{bmatrix} \begin{bmatrix} n_o\alpha_o \\ y_o \end{bmatrix} \tag{3.36}$$

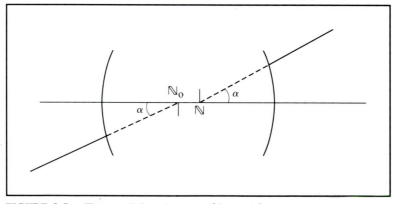

FIGURE 3.5. The nodal points are $\mathbb{N}_o$ and $\mathbb{N}$.

then $n\alpha_f = n_o\alpha_o/\mu$. From this, using the values for μ found in equations (3.29) and (3.32), one gets

$$\mathbb{N} = \frac{c - (n_o/n)}{a} \tag{3.37}$$

the image nodal point $\mathbb{N}$, and

$$\mathbb{N}_o = \frac{(n/n_o) - b}{a} \tag{3.38}$$

the object nodal point $\mathbb{N}_o$, where n_o is the refractive index of object space and n is the refractive index of image space.

Comparison of equations (3.29) and (3.32) with (3.37) and (3.38) shows that these equations are identical except for the ratio of the refractive indices that appears in each of the nodal point equations. When the refractive indices of object space and image space are the same, the nodal points are at the poles of the principal planes.

As defined in Chapter 2 with the traditional lens equation, the *focal points* are points conjugate with infinity. The image focal point is conjugate with the point at infinity in object space, and the object focal point is conjugate with the point at infinity in image space.

If one examines the a term in (3.16)

$$a = \frac{n' - n}{R} \tag{3.39}$$

one can see that it is the surface power ϕ as defined in equation (2.16). If one takes the lens shown in Figure 3.6 with surface powers of ϕ_1 and ϕ_2 and thickness t, the system matrix

$$[\mathbf{S}] = \begin{bmatrix} 1 & -\phi_2 \\ 0 & 1 \end{bmatrix} \begin{bmatrix} 1 & 0 \\ \dfrac{t}{n} & 1 \end{bmatrix} \begin{bmatrix} 1 & -\phi_1 \\ 0 & 1 \end{bmatrix} \tag{3.40}$$

and

$$[\mathbf{S}] = \begin{bmatrix} 1 - \phi_2\dfrac{t}{n} & -\phi_1 - \phi_2 + \phi_1\phi_2\dfrac{t}{n} \\[2mm] \dfrac{t}{n} & 1 - \phi_1\dfrac{t}{n} \end{bmatrix} \tag{3.41}$$

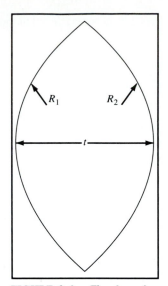

FIGURE 3.6.   The lens has surface powers ϕ_1 and ϕ_2 and thickness t.

and the a element

$$a = \phi_1 + \phi_2 - \phi_1\phi_2\frac{t}{n} \tag{3.42}$$

is, by comparison with the lensmaker's equation (2.22), the power of the system. Thus the a element of the matrix is the system's power

$$a = \phi_s \tag{3.43}$$

and

$$\frac{1}{a} = f \tag{3.44}$$

where f is the *effective focal length*, which is the name given to the focal length derived from the power of the system.

If one writes equation (3.27) in the form

$$s = \frac{\dfrac{d}{s_o} + c}{\dfrac{b}{s_o} + a} \tag{3.45}$$

and allows $s_o \rightarrow -\infty$, thereby fixing s as conjugate with $-\infty$, one gets

$$\mathbf{s} = f_b = \frac{c}{a} \tag{3.46}$$

the focal length measured from the last vertex of the system, which is then called the *back focal length*, bfl. This clearly differs from the focal length in equation (3.44) in that it has the c matrix element in the numerator. However, if one adds the bfl to the image principal plane, one finds

$$\frac{c}{a} + \frac{1-c}{a} = \frac{1}{a} \tag{3.47}$$

that is, *the effective focal length is measured from the corresponding principal plane.* It is easy to show that the object focal length is $-1/a$ and is measured from the object principal plane.

EXAMPLE 3.4   Find the cardinal points of the lens of Example 3.1 where the lens is in air.

Solution   The matrix elements of the lens are given by

$$a = 0.0758$$
$$b = 0.9824$$
$$c = 0.8052$$
$$d = -2.7594$$

The unit planes are given by

$$\mathbf{u_o} = \frac{u_o}{n} = \frac{1 - 0.9824}{0.0758}$$

$$= 0.2322$$

and $u_o = 1.00 \times 0.2322 = 0.2322$

$$\mathbf{u} = \frac{0.8052 - 1}{0.0758}$$

$$= -2.5699$$

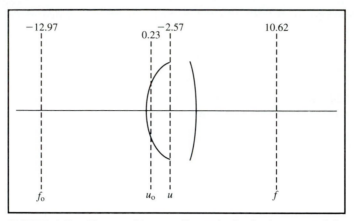

FIGURE 3.7. The positions of the principal planes and the foci in Example 3.6.

and, as with u_o, $u = -2.5699$. Since the medium of both object and image space has the same value, here 1.000, the nodal points fall at the poles of the unit planes.

The effective focal length is given by

$$f = \frac{1}{a} = \frac{1.00}{0.0758} = 13.1926$$

and $f_o = -1/a = -13.1926$.

The bfl is given by

$$\text{bfl} = \frac{c}{a} = u + f$$

and, using the results just derived,

$$\text{bfl} = \frac{0.8502}{0.0758} = 10.6227 = 13.1926 + (-2.5699)$$

The results are shown in Figure 3.7.

Mirrors

The refraction matrix for a single surface system has the form

$$[\mathbf{S}] = \begin{bmatrix} 1 & \dfrac{n' - n}{R} \\ 0 & 1 \end{bmatrix} \tag{3.48}$$

and if the elements of this matrix, namely,

$$a = \frac{n' - n}{R}$$

$$b = c = 1$$

$$d = 0$$

are placed in the imaging equation (3.27), one finds

$$\mathbf{s} = \frac{\mathbf{s}_o}{1 + \dfrac{n' - n}{R}\mathbf{s}_o} \tag{3.49}$$

This expression can be rearranged into the form

$$-\frac{1}{\mathbf{s}_o} + \frac{1}{\mathbf{s}} = \frac{n' - n}{R} \tag{3.50}$$

This equation when compared with the mirror equation (2.7) implies that one should in some way relate $(n' - n)/R$ with $2/R$. The only way in which this would be possible would be if the medium into which the ray were moving after the surface had a refractive index which was the negative of the refractive index of the incident medium. In that case, remembering that $\mathbf{s}_o$ and $\mathbf{s}$ are weighted by the index of refraction, one gets an equation identical with (2.7). Then, to incorporate mirrors into this ray-tracing procedure, we will adopt the convention that *rays traveling in the backward direction,* that is, opposite to the normal left-to-right conventional sense, *will be considered to be traveling in a medium with negative refractive index.* This is not meant to imply that there are negative refractive indices, but this is simply a procedural method to account for the directional reversal of rays when these rays are reflected. The matrix in this case will be called a *reflection matrix.*

The plane mirror with its infinite radius of curvature is an interesting case. Its system matrix

$$[\mathbf{S}] = \begin{bmatrix} 1 & 0 \\ 0 & 1 \end{bmatrix} \tag{3.51}$$

is simply the unit matrix. If one examines the image due to an object 10 cm in front of a plane mirror, that is, where $s_o = 10$ cm, then the imaging expression (3.27) yields

$$\mathbf{s} = \frac{0 + \mathbf{s}_o}{1 + 0 \cdot \mathbf{s}_o} = \mathbf{s}_o \tag{3.52}$$

One must remember that $\mathbf{s}$ and $\mathbf{s}_o$ are weighted indices and that the weighting of $\mathbf{s}$ is -1.

One then has

$$\frac{s}{-1} = \frac{s_o}{1}$$

or $s = s_o$ or 10 cm in this case. One can see then how the formalism when properly applied gives the expected result.

**EXAMPLE
3.5**

An object is 10 cm before a concave spherical mirror with radius 10 cm. Locate and describe the image. Assume that the system is in air.

Solution

The system matrix is a reflection matrix and it has the form

$$[\mathbf{S}] = \begin{bmatrix} 1 & -\dfrac{-1-1}{-R} \\ 0 & 1 \end{bmatrix}$$

$$= \begin{bmatrix} 1 & -\dfrac{2}{R} \\ 0 & 1 \end{bmatrix}$$

and $a = 2/R$, $b = c = 1$, and $d = 0$. Using equation (3.27)

$$\frac{s}{-1} = \frac{0 + \dfrac{s_o}{1}}{1 + \dfrac{-2}{-R}\left(\dfrac{-s_o}{1}\right)}$$

$$= \frac{0 + -10}{1 + \dfrac{-2}{-10}(-10)} = +10$$

and $s = -10$ cm. The image is at the position of the object. The magnification using (3.31) gives

$$\mu = \frac{1}{1 + \dfrac{(-2)}{(-10)}(-10)} = -1$$

so that the image if inverted and the same size. This result is easily verified by graphical construction as in Figure 3.8.

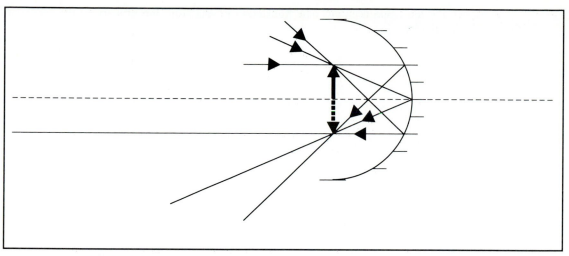

FIGURE 3.8.

An interesting example of the use of mirrors in an optical system is the Cassegrainian telescope. This is a two-mirror system as shown in Figure 3.9. The primary mirror with radius R_p forms its focus at f_p. A secondary mirror is set inside f_p to give a real image at the system's focal position f_s. In practice the mirrors are conics, but we will treat them here as spheres. The final image is at f_s. R_s is the radius of the secondary mirror.

The system matrix for the primary mirror is given by

$$\begin{bmatrix} 1 & +\dfrac{2}{R_p} \\ 0 & 1 \end{bmatrix}$$

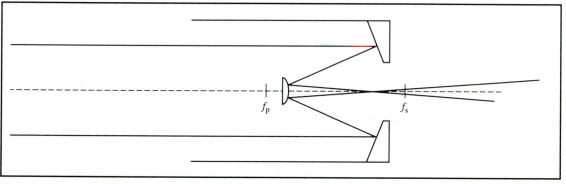

FIGURE 3.9.

since the telescope is in air. The translation matrix from the primary to the secondary mirror is given by

$$\begin{bmatrix} 1 & 0 \\ \dfrac{-D}{-1} & 1 \end{bmatrix}$$

where the ray is traveling in a negative direction in a medium of index -1 and D is the distance to the secondary mirror. The secondary mirror has a reflection matrix

$$\begin{bmatrix} 1 & -\dfrac{1-(-1)}{R_s} \\ 0 & 1 \end{bmatrix} = \begin{bmatrix} 1 & -\dfrac{2}{R_s} \\ 0 & 1 \end{bmatrix}$$

The system matrix is given by

$$[S] = \begin{bmatrix} 1 - \dfrac{2D}{R_s} & \dfrac{2}{R_p} - \dfrac{2}{R_s} - \dfrac{4D}{R_s R_p} \\ D & 1 + \dfrac{2D}{R_p} \end{bmatrix}$$

or, in terms of the focal lengths of the mirrors,

$$[S] = \begin{bmatrix} 1 - \dfrac{D}{f_s} & \dfrac{1}{f_p} - \dfrac{1}{f_s} - \dfrac{D}{f_p f_s} \\ D & 1 + \dfrac{D}{f_p} \end{bmatrix}$$

The system should have a positive focal length about equal to the focal length of the primary mirror so that the final image is positioned as shown in the figure. The back focal length of the system should be about equal to f_p. The matrix elements needed are

$$a = \frac{f_p - f_s + D}{f_p f_s}$$

$$c = \frac{D + f_p}{f_p}$$

and

$$\text{bfl} = f_p = \frac{f_s(D + f_p)}{f_p - f_s + D}$$

or

$$f_s = \frac{f_p^2 + f_p D}{D + 2f_p}$$

For a -1500-cm primary focal length and an image plane behind the primary, D should be less than 1500 cm, say, 1200 cm, in which case f_s is -2500 cm. The secondary mirror's focal length is expected to be negative since in the configuration of the telescope its radius is negative and $f_s = 2/R_s$, and the sign of the focal length is changed since the directions of the light are reversed.

Mirror systems are used extensively in the infrared long-wavelength regions, since materials which are transparent in that region have very high indices of refraction. The reflectivity of a surface is dependent upon the change in refractive index at the surface so that lenses with high indices are not very efficient in transmission.

Star Space Systems

Consider the system matrix [S] made up of two components separated by a distance t:

$$[S] = \begin{bmatrix} 1 & -\phi_2 \\ 0 & 1 \end{bmatrix} \begin{bmatrix} 1 & 0 \\ t & 1 \end{bmatrix} \begin{bmatrix} 1 & -\phi_1 \\ 0 & 1 \end{bmatrix} \tag{3.53}$$

and

$$[S] = \begin{bmatrix} 1 - \phi_2 t & -\phi_1 - \phi_2 + \phi_1\phi_2 t \\ t & 1 - \phi_1 t \end{bmatrix} \tag{3.54}$$

where the two lenses have been treated as thin and the lenses are in air so that $\mathbf{t} = t$. Now let $a \rightarrow 0$ and one finds

$$\phi_1 + \phi_2 - \phi_1\phi_2 t = 0 \tag{3.55}$$

and

$$t = \frac{1}{\phi_1} + \frac{1}{\phi_2} = f_1 + f_2 \tag{3.56}$$

so that the lenses are separated by the sum of their focal lengths. Such a system is *afocal* or *telescopic* and such systems are called *star space systems*. The system matrix for this example then has the form

$$\begin{bmatrix} 1 - \phi_2 t & 0 \\ t & 1 - \phi_1 t \end{bmatrix} \tag{3.57}$$

and has the characteristic form of all star space systems, namely, an a matrix element that is zero.

Systems such as those described by the matrix in equation (3.57) are used typically for viewing distant objects. In examining the ray-tracing equations

$$\begin{bmatrix} \alpha_f \\ y_f \end{bmatrix} = \begin{bmatrix} 1 - \phi_2 t & 0 \\ t & 1 - \phi_1 t \end{bmatrix} \begin{bmatrix} \alpha_i \\ y_i \end{bmatrix} \tag{3.58}$$

one finds

$$\alpha_f = (1 - \phi_2 t)\alpha_i \tag{3.59}$$

that is, the final angle of the ray is fixed by the value of t, the sum of the focal lengths of the two lenses. The angle of the rays leaving the system are independent of the height at which they enter the system, as illustrated in Figure 3.10.

What this amounts to is an angular magnification by the system defined by β, where

$$\beta = \frac{\alpha_f}{\alpha_i} \tag{3.60}$$

The spatial conjugates of such a system are both at infinity as can be seen from the imaging equation (3.27) with $a = 0$:

$$\mathbf{s} = \frac{d + c\mathbf{s}_o}{b} \tag{3.61}$$

If $\mathbf{s}_o \to -\infty$, then $\mathbf{s}$ will go to $\pm\infty$ depending on the sign of c. The effect is the same, however; incoming parallel rays are mapped onto outgoing parallel rays, and both object and image appear to be at infinity.

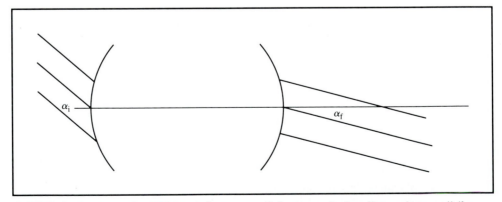

FIGURE 3.10. In a star-space system, parallel rays entering the system exit the system as a parallel bundle of rays. The angle of entry α_i is transformed into $\alpha_f = \beta\alpha_i$.

The angular magnification β of a telescope is often stated as telescopic power using the notation 7× to indicate a seven times magnification.

EXAMPLE 3.6

The terrestrial or Galilean telescope has a positive objective lens L_1 with focal length f_1 and a negative eyepiece lens L_2 with focal length $-f_2$. The separation of the lenses $t = f_1 + f_2$. Show that an erect object leads to an erect image with this system.

SOLUTION The system matrix for this system using equation (3.52) is

$$\begin{bmatrix} 1 - \dfrac{f_1 - f_2}{f_1} & 0 \\[2ex] f_1 - f_2 & 1 - \dfrac{f_1 - f_2}{-f_2} \end{bmatrix}$$

or

$$\begin{bmatrix} \dfrac{f_2}{f_1} & 0 \\[2ex] f_1 - f_2 & \dfrac{f_1}{f_2} \end{bmatrix}$$

and

$$\frac{\alpha_f}{\alpha_i} = \beta = \frac{|f_2|}{f_1}$$

This result, with β positive, implies that no inversion of the angle takes place in the telescope and erect objects yield erect images. Note that the angular magnification of this telescope is given simply by the absolute value of the ratio of the eyepiece focal length f_1 to the objective focal length f_2.

Stops and Pupils: The $f/$ Number

At this point one should have an understanding of how light passes through an optical system and forms an image. There is, however, another consideration that is of great importance with any optical system, and that concerns how much light passes through the system. Will a lens system transmit enough light to expose a film at the image plane in a reasonable period? Will a telescope pass sufficient light so that a very faint star can be seen? In terms of the optical systems that have been presented previously, these questions involve asking about the physical size of lenses in a system, not just the thickness and curvatures of the elements of the system. One cannot make a lens with a 3-cm diameter if one of its radii of curvature is 1 cm. Likewise, it is not sound practice to make a lens with a diameter of 5 cm if it is part of a system in which only its central 1 cm will be used.

The quantity of light that will pass through an optical system will be determined by the smallest clear aperture in the system. For a point on the optic axis at the design distance, that is, at the object distance for which the system is designed to function optimally, the smallest clear aperture of the system is called the *aperture stop*. Aperture stops are an essential part of an optical design, and they are set by either the diameter of one of the lenses of the system or by a diaphragm purposely inserted in the system for that purpose.

One needs first to be able to identify the element in a system that serves as the aperture stop. This might be done by tracing a marginal ray through the edge of the first lens and evaluating the *y* value of the ray at each of the lenses and diaphragms in the system. This will relate the diameter of the first lens to the remainder of the elements of the system, but if the aperture stop is not the first element of the system, all that this will fix is which stops and lenses will not interrupt the ray. If one wants to know at what height the marginal ray must enter the first lens in order to pass the system at the edge of the aperture stop, one would have to pick a new *y* value at the first surface and look for the ray height at all the other limiting apertures of the system, then iterate this process until the limiting ray is found and the stop is established.

Fortunately, a much more effective approach exists, and using this technique can be a simple and direct part of developing the system matrix for some given system. One begins by noting that the marginal ray from the edge of an object passes through the edge of its image as shown in Figure 3.11. If the marginal ray from the design distance were to pass through the edge of the second lens in Figure 3.3, one could use its conjugate image through the first lens in object space and examine the ray from the axial design point through the edge of the image of L_2 in object space. Figure 3.12 shows graphically how this is done. L_2 is imaged in L_1 and its image is located in object space. To do this

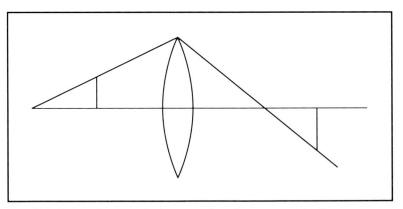

FIGURE 3.11. Conjugate images showing a marginal ray.

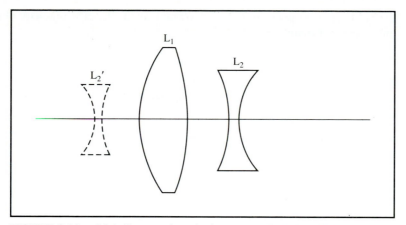

FIGURE 3.12. L_2' is the conjugate image of L_2 taken through L_1.

analytically one should note that equation (3.26) can be solved for $\mathbf{s}_o$, the object-space conjugate of an image-space quantity in terms of s, the image distance

$$\mathbf{s}_o = \frac{b\mathbf{s} - d}{c - a\mathbf{s}} \tag{3.62}$$

This will locate the object-space conjugate of L_2, and one can use the magnification equations (3.29) to find the size of the object-space conjugate.

This procedure is iterated in a multiple lens system. As each lens is added to the system matrix, the next element is projected into object space and the aperture size is determined for that lens. When all the lenses and diaphragms are projected into object space, the limiting aperture is established as the element that subtends the smallest angle at the axial point at the design distance. An example will help to clarify the procedure.

EXAMPLE 3.7

Find the limiting aperture of the Cooke triplet in Figure 3.3.

Solution

The system matrix of the first lens of the triplet has been found in Example 3.1. The first lens L_1 serves as it own aperture. The object-space aperture for L_2 can be found by projecting L_2 back through L_1. The distance $\mathbf{s} = s$ to L_2 is taken as $1.25 + 1.55/2 = 2.025$ mm, that is, its aperture is taken at its center. Using the system matrix for L_1 found in Example 3.1 and (3.62), one gets

$$\mathbf{s}_o = \frac{(0.9824 \times 2.025) + 2.7594}{0.8052 - (0.0758 \times 2.025)} = 7.2867$$

The object-space conjugate of L_2 is 7.2867 mm beyond L_1. The diameter of this conjugate is the magnified diameter of L_2:

$$D = 2.36 \times \left(\frac{1}{c - a\mathbf{s}} \right) = \frac{2.36}{0.8052 - (0.0758 \times 2.025)} = 3.62 \text{ mm}$$

To find the object-space conjugate of L_3, one needs first to find the matrix for the L_1–L_2 combination. The system matrix for L_2 can be found from the parameters given in Figure 3.3 and these yield

$$\begin{bmatrix} 1.0583 & 0.0951 \\ 0.914 & 1.0271 \end{bmatrix}$$

and for the combination L_1–S–L_2 where S is the L_1–L_2 spacing one gets

$$\begin{bmatrix} 1.4189 & -0.0127 \\ 4.9939 & 0.6604 \end{bmatrix}$$

The object-space conjugate of L_3 with this system matrix is

$$\mathbf{s}_o = \frac{(1.4189 \times 5.00) + 4.9939}{0.6604 + (0.0127 \times 5.00)} = 20.26 \text{ mm}$$

The object-space conjugate of L_3 is 20.26 mm behind the first vertex of the L_1–L_2 combination, well beyond the system itself. This is not surprising since the L_1–L_2 combination is a very weak lens. Figure 3.13 gives the positions of the poles of the system and the object space conjugates of L_1, L_2, and L_3. The size of the conjugate of L_3 is given by

$$D = 2.58 \times 119.2 = 307.6 \text{ mm}$$

Clearly the limiting aperture is that formed by L_1.

The limiting entrance aperture in object space is called the *entrance pupil*. The entrance pupil thus fixes the amount of light that will be accepted by a lens system. When the entrance pupil is imaged through the entire lens system, its image is the *exit pupil*. The exit pupil fixes the size of the light cone converging to form the image.

While a knowledge of the entrance and exit pupils is important in designing and specifying lens systems, one needs a method for easily describing the light gathering properties of a lens. Such a description is usually given by the *f*-number, *f*/, which is the ratio of the effective focal length of the system to the clear aperture of the system

$$f/ = \frac{f}{d_{cl}} \tag{3.63}$$

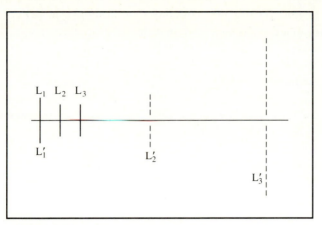

FIGURE 3.13. The conjugates of L_1, L_2, and L_3 referred to the poles of the triplet P_1 and P_2. The diameters of the images are shown.

EXAMPLE 3.8   What is the f-number of the triplet lens in Figure 3.3?

Solution   The system matrix of the triplet is given by

$$\begin{bmatrix} 0.5484 & -0.0592 \\ 11.8140 & 0.5469 \end{bmatrix}$$

The focal length of this system is $1/a$ or 16.86 mm, and the f-number ($f/$) is found using the first lens as the limiting aperture using equation (3.62)

$$f/ = \frac{16.86}{3.00} = 5.62$$

Note that the larger the f-number, the slower the lens, that is, the less light gathering capability the lens possesses.

PROBLEMS

3.1. Find the complete system matrix for the doublet in Table 3.1. Note that the system matrix is made up of two simple lenses separated by an air space. The system matrix of each lens can be found first and then combined. Demonstrate that this is a valid approach.

3.2. Verify the system matrix for the Cooke triplet specified in Figure 3.3.

3.3. (a) Find the cardinal points for the lens in Problem 3.1. (b) Find the cardinal points for the Cooke triplet in Problem 3.2 and verify that the bfl is equal to the sum of the effective focal length and the image principal plane.

3.4. Examine the effect of a thin lens of arbitrary power placed at the object focal plane of a lens system.

3.5. Find the front focal length, ffl, of a lens system, that is, the object-side focal length measured from the first vertex of the system.

3.6. Find the inverse of the system matrix and show that the result $s_o = f[s]$ which one gets using it is the same as one finds with the direct matrix.

3.7. The Cooke triplet in Figure 3.3 is to have an f-number of 3.5. What is the new diameter of the first lens?

3.8. Given a lens that is defined as thin in the paraxial model, that is, $t \to 0$. Show that the principal planes are coincident with the plane of the lens.

3.9. The radius of curvature of a concave spherical mirror is 50 cm. An object 3 cm tall is located before the mirror at distances (a) 60 cm, (b) 30 cm, (c) 15 cm, and (d) 5 cm. Locate and describe the images in each case.

3.10. A thin lens with index 1.5 is symmetric and positive with radii 12 cm. One face of the lens is silvered. Find the system matrix for light entering through the unsilvered face.

3.11. The curved surface of a planoconvex lens is silvered. Discuss the optical properties of this lens.

3.12. A glass sphere with index n_o is silvered in one hemisphere (Figure P3.12). Light enters the sphere through the unsilvered hemisphere. Find the system matrix.

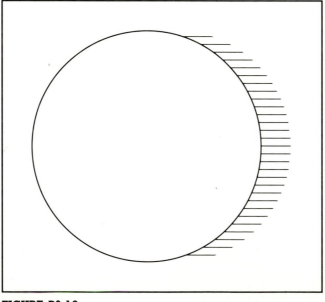

FIGURE P3.12.

3.13. Investigate a Cassegranian-type telescope where $D > f_p$. Such a telescope is called a Grego-
rian telescope. Specifically, what is the sign of f_s?

3.14. (a) Consider a telescope where both the eyepiece and the objective are positive. Find an
expression for the magnification of the system in terms of the focal lengths of the elements. Is
the image erect or inverted? (b) The telescope inverts the image. The system can be made
erecting by inserting an erecting element between the objective and the eyepiece. The erecting element operates between the foci of the two original lenses with a magnification of -1.
Sketch the system and show that it is telescopic and that the angular magnification is not
dependent on the erecting element.

3.15. How would one fix the size of the eyepiece of a Galilean telescope for a given size objective?

3.16. A telescopic system is to be used with its objective lens set in the side of a submarine for
underwater viewing. How does the presence of the water medium in object-space affect the
angular magnification of the system?

3.17. Adjust the diameter of L_1 in Example 3.7 so that it remains the limiting aperture for a distant
object while getting better utilization of L_2 and L_3. What is the change in the f-number, if
any, resulting from this change?

3.18. Show that Newton's lens equation can be used with a general optical system such as those
treated in this chapter. What focal length is used?

3.19. Show that the thin-lens graphical construction that takes the line from the tip of the object to
the tip of the image through the center of the lens *without an alteration in direction* is valid.

CHAPTER 4

Designing Lenses

The previous two chapters established the basis of the geometrical model and, in particular, illustrated how this model can be used to describe imaging in lenses and mirror systems. The various imaging relationships that were found were restricted in that they are based on the paraxial assumptions. In this chapter some issues in the design of optical systems are discussed. These include the choice of materials as well as the result of the relaxation of the paraxial approximation and the use of the more precise trigonometric ray tracing.

Relaxation of the Paraxial Approximation

The paraxial approximation as introduced in Chapter 2 involved replacing the tangent or sine of an angle with the angle itself, resulting in Snell's law (2.1) taking the form

$$n\theta = n'\theta' \tag{4.1}$$

and in the assumption that the sagittal depth of a surface, the axial distance from the edge of the surface to its intersection with the optic axis, is vanishingly small. This is equivalent to a scaling of the aperture of the system by a vanishingly small factor, say, 10^{-10}. What happens as this scaling factor is relaxed and unscaled apertures are examined?

Equation (4.1) can be gotten by taking only the first term in the series expansion for $\sin \theta$:

$$\sin \theta = \theta - \frac{\theta^3}{3!} + \frac{\theta^5}{5!} - \frac{\theta^7}{7!} + \cdots \tag{4.2}$$

in Snell's law. The terms can be introduced one at a time so that the first improved approximation is, for example,

$$n\left[\theta - \frac{\theta^3}{3!}\right] = n'\left[\theta' - \frac{\theta^3}{3!}\right] \tag{4.3}$$

This was one path by which the design of optical systems evolved. The deviations from the paraxial predictions are called *aberrations*. When only the cubic term of the sine expansion is included the aberrations are called third-order aberrations. Those including the fifth-power angle term in (4.2) are called fifth-order aberrations, and so on. With the third-order approximation in equation (4.3) and with monochromatic light the aberrations were studied by Seidel and were divided into five classes:

1. Spherical aberration
2. Coma
3. Astigmatism
4. Field curvature
5. Distortion

There is an additional aberration, chromatic aberration, resulting from the relaxation of the monochromaticity requirement, and this will be treated separately.

Traditional design involved using expression (4.3) or used additional higher-order terms to evaluate the aberrations, followed by adjustment of the parameters of the system to reduce the aberrations to some acceptable minimum. It will be shown later when the wave model is developed that it is impossible to have a point image for a point object in the general case; there is some minimum value beyond which refinement of the system has no value.

Today lens design is most often approached using very highly developed computer programs, but the use of the terminology and ideas of the Seidel aberrations has been retained and remains part of the design process. After the paraxial layout of the system has been made, calculation of the third-order aberrations is typically the next step in the design process.

Optical Materials

The system layout, the first design step, is done using the specifications of the system, focal length, back focal length, aperture, and so on. A significant element in this process is the initial selection of the glasses or other materials for the lens elements. There is a remarkably large range of glasses available to the designer, ranging in price from pennies to tens of dollars per pound. Figure 4.1 is a *glass map*, a plot of the refractive index of the glasses available from Schott Glass Technologies, Inc., versus a measure of the change of refractive index with wavelength, denoted V. This glass map is taken from the Schott catalog. The vertical line at $V = 50$ is traditionally taken as the line separating crown

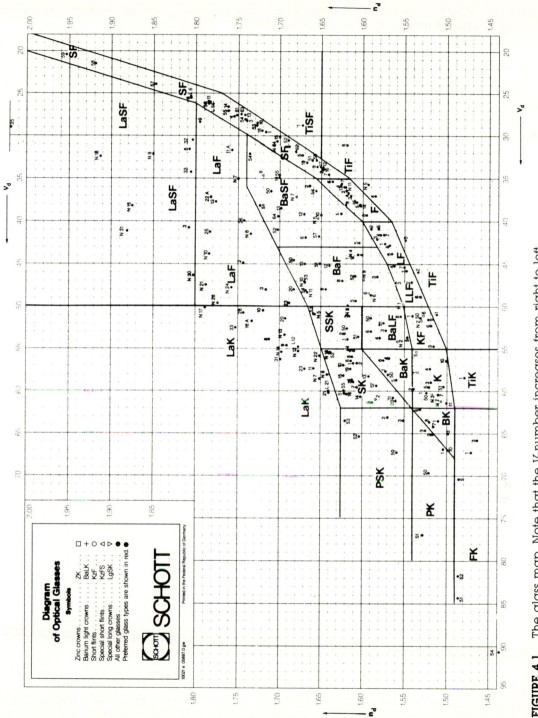

FIGURE 4.1. The glass map. Note that the V-number increases from right to left.

glasses (German, *kron*) with a *K* designator from the flint glasses with an *F* designator. Part of the selection process for glasses involves the *V* number, a measure of the dispersive property of the glass, which will be discussed in the section dealing with chromatic aberration.

The parabolic-shaped arc with the dense cluster of glasses running from the lower left to the upper right of the glass map is called the *glass line*. The original optical glasses were scattered along this line. The addition of other oxides to the glasses, such as barium and strontium as well as some rare earth materials, has generated many additional optical glasses placed above the glass line.

Glass is provided in several forms, including slabs, blocks, strips, rods, and pressings. When ordering glass one should specify not only the refractive index and *V* number, that is, the glass type, but also a number of physical and chemical properties. Homogeneity, including the presence of striae, threadlike inclusions whose index differs slightly from the base glass but which are visible imperfections, must be considered. Resistance to chemicals such as acids and bases as well as to weathering may be important depending on the application. In certain environments, thermal properties, especially expansion, become important. Bubble content of the glass clearly is another property important in optical systems; the bubbles in a glass bottle generally do not affect the function of the glass, but bubbles in a lens are a serious detriment.

Besides glasses, there are many other materials such as plastics that are transparent in the visible region of the spectrum and that possess sufficient stability to be used as lenses. These materials are used extensively in spectacles and contact lenses, for example. For the infrared region there is a wide selection of crystalline materials, including the semiconductors silicon and germanium and the alkali halides.

Spherical Aberration

The first of the monochromatic aberrations in the Seidel list is the only aberration that occurs for object points on the optic axis. As with all the low-order Seidel aberrations, there are techniques that have been developed to find these aberrations from the paraxial parameters, and these are discussed in Smith (1990). Here no attempt will be made to establish these rules since a more general approach to the calculation of the imaging errors will be given and related to the computer-based design techniques.

Spherical aberration arises from the difference in focal position for rays through the margins or edges of the lens compared with those paraxial rays near the axis. Figure 4.2 illustrates this point. Each surface in the lens system makes a contribution to the spherical aberration, and the aberration is assigned a value by giving the difference between the marginal focus *L* and the paraxial focus *l*, which is called the *longitudinal spherical aberration*, LSA:

$$LSA = L - l \tag{4.4}$$

The LSA may be either positive or negative depending on whether the surfaces are undercorrected (LSA positive) or overcorrected (LSA negative).

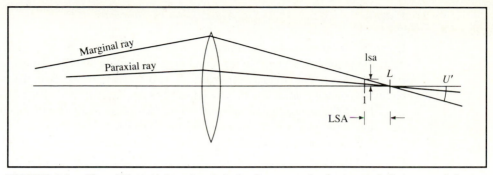

FIGURE 4.2. The different focal points for the marginal ray and the paraxial ray give rise to the spherical aberration.

A second measure of the spherical aberration is the lateral spherical aberration, lsa, the height of the marginal ray at the paraxial focal point. This can be calculated easily when one knows the LSA and the angle between the marginal ray and the optic axis, U'.

$$\text{lsa} = (\text{LSA}) \tan U' \tag{4.5}$$

Actually, the best focus is usually not at either the paraxial focus or the marginal focus but somewhere between them. Figure 4.3 shows the rays in the region of the foci. As can be seen from the figure, the minimum size of the image spot is at the point where the marginal rays intersect the paraxial rays. This spot is typically between the marginal and paraxial foci and is called the *circle of least confusion*. Most lenses are well corrected for LSA, and the LSA will often be of the order of microns or several wavelengths when the lens is used at its design distance.

The transverse third-order spherical aberration depends on the square of the aperture of the lens. It is possible, however, to reduce the LSA by properly shaping the lens, as will be seen in the next section.

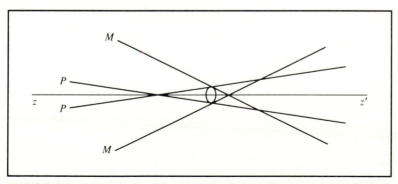

FIGURE 4.3. The circle of least confusion as shown occurs at the intersection of the marginal and paraxial rays.

Coma

Coma is the first of the aberrations that arise with off-axis sources. While spherical aberration remains, with off-axis sources the other four Seidel aberrations take on increasing importance. The oblique refraction of the off-axis rays results in the breaking of the circular symmetry present with on-axis sources. As a result two classes of rays must be considered: the *meridional or tangential rays,* which lie in a plane that contains the optic axis; and *skew rays,* which enter the optical system with an x coordinate other than zero and usually do not intersect the optic axis of the system. One special ray used in ray tracing is the *chief ray*, which is the ray from the object passing through the center of the entrance pupil. The set of skew rays that lies in the plane containing the chief ray and is normal to the optic axis is given the special name *sagittal rays.*

Figure 4.4a illustrates the formation of the comatic image in the meridional plane. The loss of symmetry with the off-axis object results in the spreading of the image in the image plane. Figure 4.4b shows the form of the image where the circles can be thought of as arising from different annuli of the lens. Note that while circles are used in the illustration, the image itself is continuous, and the name ''coma'' arises from the appearance of the image in the shape of a comet. Figure 4.5 is the ''comatic clock,'' which further illustrates the formation of coma. The points labeled 1 in the circle on the lens combine at point 1 in the image, as do points 2, 3, and 4.

Coma is particularly deleterious in cameras with their large fields of view since the comatic patches increase in size toward the periphery of the image. It is of less concern in systems where the image lies close to the optic axis such as in microscopes.

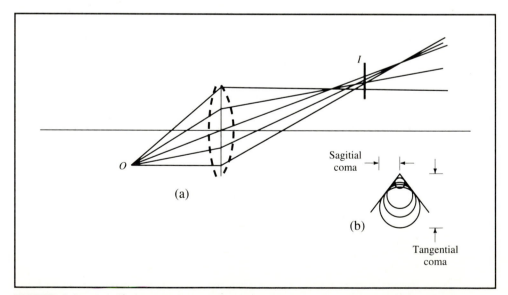

FIGURE 4.4. (a) The formation of the comatic image in the meridional plane. (b) The appearance of the comatic image where the circular patches are due to annuli of the lens.

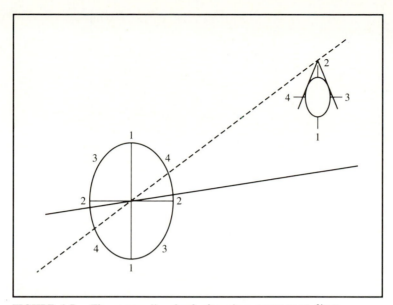

FIGURE 4.5. The comatic clock showing corresponding positions on the lens and on the image.

Like spherical aberration, coma depends on the square of the radius of the aperture. It can be made to vanish by properly shaping the lens. The lensmaker's equation (2.22) relates the power of a lens to the two surface radii. Once the power and one radius are selected, the second radius is then fixed. To examine the elimination of LSA and coma, one begins by defining the shape factor for a lens,

$$q = \frac{r_2 + r_1}{r_2 - r_1} \qquad (4.6)$$

Figure 4.6 gives the general shape of several lenses and their shape factors. Note that the shape factor of the symmetrical biconvex lens is 0 and that the shape factor is continuous over an extended range.

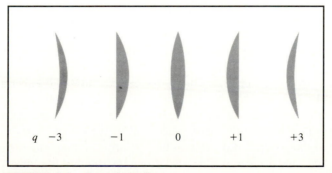

FIGURE 4.6. Lens bending.

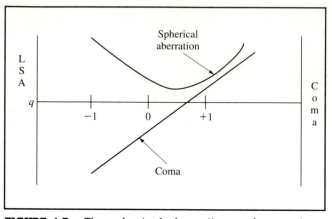

FIGURE 4.7. The spherical aberration and coma for a positive lens as a function of the lens bending. The LSA minimum lies to the left of the coma zero.

If one plots third-order spherical aberration and coma against the shape factor for a single lens element, one finds the result shown in Figure 4.7. Third-order spherical aberration has a minimum value for some shape factor for a positive lens in air at a given design distance while the coma has a zero value for a slightly higher shape factor. The general result is that the shaping of the lens element at a proper value of q has the effect of correcting the LSA and the coma simultaneously.

Systems with no LSA or coma are called *aplanatic systems*. Only in very special cases can this be done with a spherical single lens element. Generally, when an aplanatic single lens element is required one is forced to use aspheric surfaces, an expensive solution.

Astigmatism

Even if both LSA and coma are absent in a system, a problem may remain with the difference in focus between the tangential rays and the sagittal rays. This is illustrated in Figure 4.8. The positions of these two foci are displaced from one another, giving two lines of focus as shown in the figure, and the aberration resulting from the displacement of these two foci is called *astigmatism*. The best focus of the system lies between these two lines, and the image patch varies from the lines to an ellipse, to a circle, to an ellipse with major axis at 90° to the ellipse at the other side of the circular image, and finally back to a line at 90° to the first line.

Astigmatism vanishes for objects on the optical axis, like all the aberrations with the exception of spherical. As the object is displaced from the optic axis, however, the separation of the tangential and sagittal images increases. This is illustrated in Figure 4.9. The astigmatism is dependent on the displacement of the object from the optic axis as noted but also on the position of the stop of the system. The stop establishes the aperture of the system and thus the size of the bundle of rays forming the image. For a single

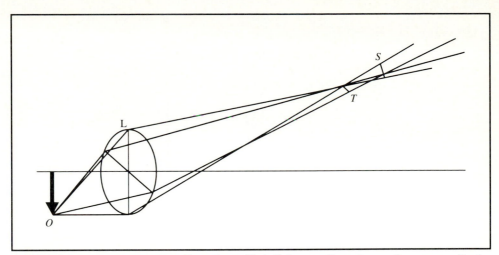

FIGURE 4.8. The tangential, *T*, and sagittal, *S*, image lines formed as a result of astigmatism in lens L for an object O.

element, the astigmatism is dependent on the square of the image height and inversely on the focal length of the element; the general shapes of the two foci are parabolic, as one can see from Figure 4.9.

Curvature of the Field

The previous section on astigmatism has shown that the image formed by a lens does not lie in the paraxial plane or in any other plane; rather, it lies along some paraboloid surface. Every lens system has a fundamental surface called the *Petzval surface* that is associated

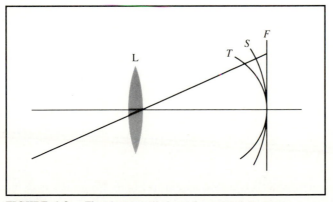

FIGURE 4.9. The tangential and sagittal image curves for a simple lens element L.

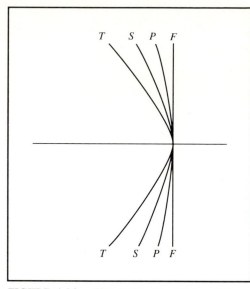

FIGURE 4.10.   The Petzval surface, *P*, is shown in relation to the paraxial focal plane, *F*, the sagittal focus, *S*, and the tangential focus, *T*, for the case of a positive lens.

with the aberration known as *field curvature*. This is illustrated in Figure 4.10. It should be noted that the Petzval surface lies between the paraxial focal plane and the astigmatic foci.

With a positive lens or lens system the Petzval curve is concave toward the lens as in Figure 4.10, while with a negative lens it is convex. The distance to the Petzval curve from the tangential image is three times the distance to the sagittal image. For a single element, the Petzval curve depends inversely on the refractive index of the element. Correction of field curvature is called *field flattening*, and it is achieved by separating the elements of a system as well as by careful choice of the glasses used in the system.

For a thin lens with index *n* the Petzval radius is given by $r_{\mathrm{p}} = -nf$, where *f* is the focal length of the lens. For a system of thin lenses

$$\frac{1}{r_{\mathrm{p}}} = -\sum_j \frac{\phi_j}{n_j} \tag{4.7}$$

If one expands this expression for a single lens element with two surfaces, one gets

$$\frac{1}{r_{\mathrm{p}}} = \sum_{j=1}^{2} \frac{(n_j - 1)}{n_j} \left(\frac{1}{R_1} - \frac{1}{R_2} \right) \tag{4.8}$$

by applying the lensmaker's equation (2.22) to find *f*. If one wants to have a flat image plane, as is often the case, then the target is a large r_{p}, which is best gotten by large radii of curvature on the lens surfaces.

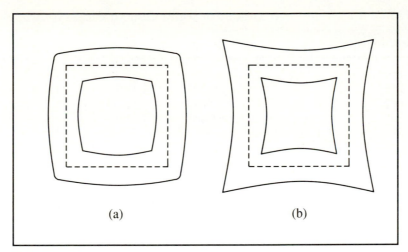

FIGURE 4.11. Distortion. (a) Barrel and (b) pincushion.

Distortion

Distortion arises whenever the magnification is dependent upon the distance of the image from the optic axis. The distortion usually has a cubic dependence on the object height, and the appearance of a square object can take on either of two typical forms shown in Figure 4.11. The left-hand section of the figure is related to barrel distortion, which arises when the magnification decreases with image size, while the right-hand section illustrates pincushion distortion, typical of the situation where the magnification increases with image size.

If one applies the principle of optical reversibility here, one quickly sees that exchanging the image and object results in the transformation of the system from barrel to pincushion, or vice versa. One approach to removing distortion in a camera lens, for example, is to make the lens symmetrical about its center where the stop is placed. Most systems will tolerate a small amount of distortion, but if the distortion reaches 7–8% image problems are often apparent.

Chromatic Aberration

Unlike the Seidel aberrations, *chromatic aberration* is, as its name implies, dependent upon the light's color or wavelength. The index of refraction of a material is wavelength dependent; that is, materials are optically dispersive. Table 4.1 lists the refractive indices of some typical optical glasses at several standard wavelengths. As can be seen from the table, the index decreases continually from the blue (F') to the red (C). The decrease is nearly linear in $1/\lambda^2$, so that $1/\lambda^2$ serves well as an interpolating factor to find n at intermediate values.

TABLE 4.1   Refractive Indices of Some Optical Glasses

Material	n_d	n_C	$n_{F'}$	n_g	n_h
K5	1.52249	1.51982	1.52910	1.53338	1.53735
LLF1	1.54814	1.55768	1.57082	1.57713	1.58313
LF5	1.58144	1.57723	1.59230	1.59964	1.60667
F6	1.63636	1.63108	1.65017	1.65963	1.66879
SF1	1.71736	1.71032	1.73610	1.74916	1.76199
λ (nm)	587.56	656.27	479.99	435.83	404.66

The V value, which is the abscissa of the glass map, is a measure of the dispersion of a particular glass and is defined as

$$V = \frac{n_d - 1}{n_F - n_C} \tag{4.9}$$

The value n_d is the index near the wavelength of the maximum sensitivity of the human eye, and when a single index of refraction is given the value is typically n_d. Note that the d-line is near but is not one of the familiar sodium yellow doublet lines.

The chromatic aberration may be defined as the change in power of a lens with refractive index. Using the thin-lens lensmaker's equation (2.22), one has

$$d\phi = dn\left(\frac{1}{R_1} + \frac{1}{R_2}\right) \tag{4.10}$$

and dividing by (2.22), one gets

$$\frac{d\phi}{\phi} = \frac{dn}{n-1} \tag{4.11}$$

or

$$d\phi = \frac{dn}{n-1}\phi \tag{4.12}$$

where ϕ is the lens power.

If one takes $dn = (n_{F'} - n_C)$, then

$$d\phi = \frac{\phi}{V} \tag{4.13}$$

which implies that any lens with nonzero power will suffer chromatic aberration.

This aberration may be corrected by recognizing that different glasses can be combined to eliminate $d\phi$. Consider a thin-lens doublet where the elements are cemented together. The power is

$$\phi = \phi_1 + \phi_2 \qquad (4.14)$$

where ϕ is the specified lens power. Now

$$d\phi = d\phi_1 + d\phi_2 \qquad (4.15)$$

and if $d\phi$ is to be zero, then

$$d\phi_1 = -d\phi_2 \qquad (4.16)$$

or

$$\frac{\phi_1}{V_1} = -\frac{\phi_2}{V_2} \qquad (4.17)$$

Combining the appropriate glasses will then yield a lens with no primary chromatic aberration. There are additional color image defects resulting from the inclusion of higher-order terms since the treatment just given is purely paraxial and is adjusted to the end points (F' and C lines) of the visible spectrum. The uncorrected colors give a *secondary chromatic aberration*.

Substituting equation (4.15) into equation (4.12) yields two equations for the powers of the thin-lens elements:

$$\phi_1 = \frac{V_1}{V_1 - V_2}\phi \qquad (4.18)$$

$$\phi_2 = \frac{V_2}{V_1 - V_2}\phi \qquad (4.19)$$

Examination of Table 4.1 or, better, the glass map of Figure 4.1 (where V increases toward the left) shows that the V numbers for flint are in the 25–50 range while those of the crown glasses are in the 50–75 range. Thus it is possible to use these glasses pairwise to get chromatic correction.

EXAMPLE 4.1

An achromatic lens is to have a power of 10 D. Give the curvatures if the crown is K5 and the flint is F6 and if the first radius of the doublet is to be 15 cm. Neglect the thicknesses. What is the Petzval radius?

Solution

From Table 4.1 $V_1 = 59.48$ and $V_2 = 35.34$, so that $V_1 - V_2 = 24.14$. Using equations (4.16) and (4.19) one finds

$$\phi_1 = \frac{59.48}{24.14}10 = 24.64 \quad D \qquad (4.20)$$

and

$$\phi_2 = -\frac{35.34}{24.14} \, 10 = -14.64 \quad D \tag{4.21}$$

Using the lensmaker's equation (2.22) with the crown lens 1, one gets

$$24.64 = (1.52249 - 1.00)\left(\frac{1}{0.15} - \frac{1}{R_2}\right) \tag{4.22}$$

or $R_2 = -2.470$ cm. Continuing with lens 2, the flint lens, where the first radius is given by R_2,

$$-14.64 = (1.63636 - 1.00)\left(\frac{1}{-0.02470} - \frac{1}{R_3}\right) \tag{4.23}$$

and $R_3 = -5.721$ cm. The negative flint lens is meniscus with this choice of R_1. The Petzval radius is found by using (4.7):

$$\frac{1}{r_p} = -\left(\frac{24.64}{1.52249} - \frac{14.64}{1.62626}\right) \tag{4.24}$$

and

$$r_p = -13.82 \quad cm \tag{4.25}$$

This example, as well as equations (4.18) and (4.19), shows that the correction of primary chromatic aberration is independent of the lens bending. The correction is also independent of the object position, and this in itself distinguishes chromatic aberration from the Seidel aberrations.

Trigonometric Ray Tracing

In Chapter 3 paraxial rays were traced through optical systems using matrix techniques. This is useful within the constraints of the paraxial approximation, but it does not provide the detailed information needed when design is being undertaken. In this section the precise trigonometric trace procedure for meridional (tangential) rays is presented. Those interested in the more general techniques, including skew-ray tracing, should consult Kingslake (1978).

There are a number of techniques for meridional traces that have been developed over many years. Some are best suited to "hand" calculation, which was the principal

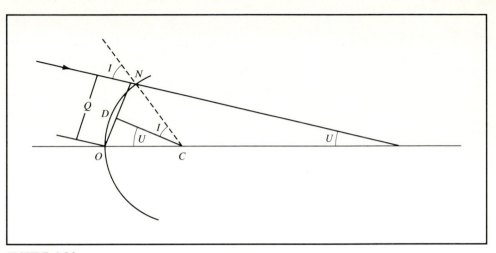

FIGURE 4.12.

method prior to the arrival of electronic calculators and computers. In fact, one of the earliest uses of computers was in ray tracing and optical design. The approach developed here is one that lends itself to the use of the calculator.

The geometry of the trace is shown in Figure 4.12. The ray is incident from the left and has an angle of incidence I with respect to the surface normal, that is, with the radius of the surface drawn to that point. The ray is specified by its intersection with the optic axis relative to the pole of the surface, O, and the angle U it makes with the optic axis. One first needs to find Q, the normal distance from the pole of the surface to the incoming ray (or its extension). To this end one draws a parallel to the ray from the center of curvature C to the normal to the ray. This divides Q into two parts, OD and DN. From the figure one has

$$Q = OD + ON \qquad (4.26)$$

and

$$Q = r \sin I + r \sin U \qquad (4.27)$$

which gives

$$\sin I = \frac{Q}{r} - \sin U \qquad (4.28)$$

Another sign convention is needed here. I is taken as positive if rotation of the surface normal into the ray is an anticlockwise rotation.

Once $\sin I$ is found, Snell's law gives I',

$$\sin I' = \frac{n}{n'} \sin I \qquad (4.29)$$

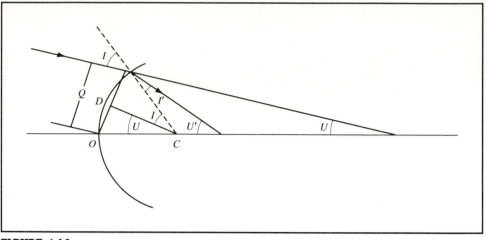

FIGURE 4.13.

After refraction, one can see from Figure 4.13 that

$$U + I = U' + I' \tag{4.30}$$

since these are both supplements of the same angle so that

$$U' = U + I - I' \tag{4.31}$$

and

$$Q' = r \sin U' + r \sin I' \tag{4.32}$$

by analogy with Q. Once U' and Q' have been found, one can find the new position of the ray's intersection with the optic axis.

In the transfer to the next surface, the lines defining Q remain parallel since they are both normal to the same ray. One can see from Figure 4.14 that

$$Q' = Q - d \sin U' \tag{4.33}$$

The procedure for a ray trace is then as follows:

1. The opening value of L and U is given, from which the value of Q, Q_{init}, is given by

$$Q_{\text{init}} = L \sin U \tag{4.34}$$

2. Equations (4.27), (4.28), (4.29), and (4.30) are used to find the new Q following refraction.

3. The transfer equation (4.32) then is used to establish the Q for the next surface and the procedure is continued.

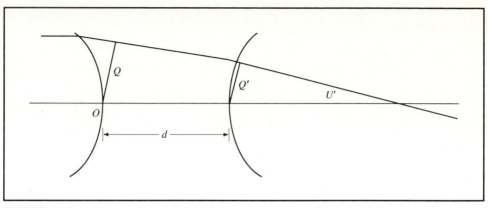

FIGURE 4.14.

4. The final L, which is the precise bfl of the system, is given by

$$L_{\text{final}} = \frac{Q'}{\sin U'} \qquad (4.35)$$

where Q' and U' are the values following refraction at the last surface of the system.

EXAMPLE 4.2   Use the results of Example 4.1 with the thicknesses specified in Table E4.2A to fix the LSA for an object at infinity.

TABLE E4.2A   Achromatic Doublet Objective[a]

Surface	1		2		3
Radius (cm)	15.00		−2.470		−5.721
n_d		1.52249		1.63636	
t (cm)		1.25		0.2	

[a] The diameter of the doublet is 4 cm.

Solution   The system matrix for the lens in Figure E4.2 is found to be

$$\begin{bmatrix} 0.932416 & -0.098236 \\ 0.947871 & 0.972617 \end{bmatrix}$$

The focal length of this lens is given by $1/a$ and is 10.179568 very close to the design target in Example 4.1. It should be noted, and this is usually the case, that the thicknesses of the elements do not strongly affect the design. The back focal length, c/a, is 9.900820 cm.

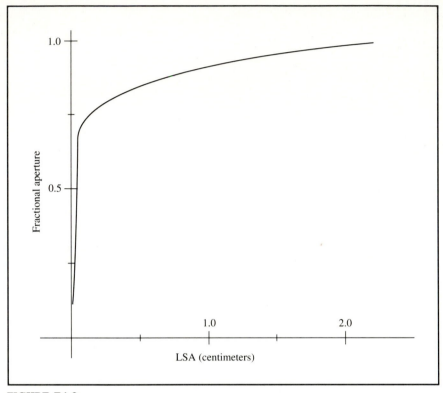

FIGURE E4.2.

The trigonometric trace is done by setting up a table as shown here and computing the individual terms using the trace equations. The object is at infinity, so that the ray entering the system comes in parallel to the optic axis and Q_{init} here will be be 2.00 cm for a marginal ray. The data are placed in Table E4.2B as generated.

TABLE E4.2B   Aperture Radius = 2.00 cm

Surface	1	2	3
Q	2.000000	1.946514	2.055780
U	0.0	2.638094	−2.982959
$\sin U$	0.0	0.046027	−0.052039
$\sin I$	0.133333	−0.834089	−0.307300
I	7.662256	−56.521119	−17.896609
$\sin I'$	0.087576	−0.776047	−0.502854
I'	5.024162	−50.900066	−30.188993
U'	2.638094	−2.982959	9.309425
$\sin U'$	0.046027	−0.052039	0.161766
Q'	2.004047	2.045372	1.951364
L_{final} = 12.062878			

Clearly the position of the axial image point for a ray through the margin of this system is very different from that found in the paraxial trace. The system has significant spherical aberration.

To examine this one traces a ray at a fractional aperture. This is initially taken at 1.4 cm or at a fractional aperture of 0.7 as shown in Table E4.2C.

TABLE E4.2C Aperture Radius = 1.4 cm

Surface	1	2	3
Q	1.400000	1.361227	1.385681
U	0.0	1.840789	−0.973382
$\sin U$	0.0	0.032122	−0.016988
$\sin I$	0.093333	−0.583226	−0.225222
I	5.355401	−35.677770	−13.015908
$\sin I'$	0.061303	−0.542641	−0.368544
I'	3.514612	−32.863599	−21.625824
U'	1.840789	−0.973382	7.636534
$\sin U'$	0.032122	−0.016988	0.132888
Q'	1.401380	1.382283	1.348186
L_{final} = 9.999998			

The value of 1.4 was chosen as an aperture, since the fractional aperture radius of 0.707 divides the lens into two equal-area portions. This fractional aperture is often taken as the aperture at which a system is to be corrected, particularly if some compromise is required. Here the spherical aberration is significantly smaller than at full aperture.

As a final example a third trace at a fractional aperture of 0.5 is given to illustrate the dependence of the spherical aberration on the aperture. This is shown in Table E4.2D. This trace would rarely be undertaken during the nonautomated design of this lens.

TABLE E4.2D Aperture Radius = 1.0 cm

Surface	1	2	3
Q	1.000000	0.971863	0.980198
U	0.0	1.312889	−0.500217
$\sin U$	0.0	0.022912	−0.008730
$\sin I$	0.066667	−0.416379	−0.162603
I	3.822554	−24.606181	−9.358038
$\sin I'$	0.043788	−0.387404	−0.266078
I'	2.509665	−22.793075	−15.420990
U'	1.312889	−0.500217	5.572735
$\sin U'$	0.022912	−0.008730	0.097109
Q'	1.000503	0.978452	0.966670
L_{final} = 9.95484			

One can see, then, that the bfl at this aperture is approaching the paraxial value. A plot of the spherical aberration is shown in the figure. Note that what is shown is not the third-order spherical aberration but rather the spherical aberration in all orders.

If the design does not give the correct focal length but is satisfactory in other respects, there is no need to redo the design completely. One can simply *scale* all the linear dimensions by the ratio of the desired focal length to the calculated focal length. The resulting lens will have the same properties and the required design focal length. This scaling is equivalent to a change in units so that the aberrations will also change by the same scaling factor. If, for example, a focal length of 12 cm was found when a 10-cm focal length was required, one would multiply all the linear dimensions of the problem (radii, thicknesses) by 10/12 or 0.8333.

Automated Lens Design

As noted, computers have had a major impact on the way lenses are designed. Today there as several very good design packages available to the lens designer that will allow him or her to perform design calculations with great rapidity. Not only can the designer make changes in a system and quickly evaluate the effect of these changes, but he or she can turn the optimization of the system over to the computer program and get a lens optimized to a set of target values for the aberrations.

In what follows the data generated are from the OSLO system as provided by Sinclair Optics. This is one of the better programs available today for small computer systems.

The data from the example lens as specified in Example 4.2, when entered into the OSLO program, can be called out using an rtg (radii, thicknesses, glasses) command to yield the data shown in Figure 4.15. There are five surfaces specified here. Surface 1 is reserved for the entrance aperture and surface 5 is the image plane. One can note that the

```
LENS DATA
Book Example
SRF     RADIUS        THICKNESS        APERTURE        GLASS
1        --              --            2.000000 A       AIR
2      15.000000     1.250000          2.000000 S       K5  C
3      -2.470000     0.200000          1.951013 S       F6  C
4      -5.721000     9.900796 S        1.954714 S       AIR
5        --             --      S      0.101795 S       AIR
```

FIGURE 4.15.

```
*GENERAL DATA
 PROG: FDU12123
      EPR             OBY            THO         CVO         CCO        UNITS
  2.000000      -1.00000E+18    1.00000E+20     --          --       1.000000
      IMS      AST     RFS     AFO     AMO     DESIGNER    IDNBR
       5        1       1       0      TRA        JWB         6
 *PARAXIAL CONSTANTS
      EFL            FNB          GIH          PIV        PTZRAD         TMAG
   10.179535     2.544884     0.101795    -0.020000    -13.821781   -1.01795E-19
```

FIGURE 4.16.

thickness of surface 4 is 9.900796, which corresponds to the bfl of the system found earlier. The glasses are entered by the catalog number and are drawn from the glass catalog that is part of the design system. It is not necessary to enter all the indices whenever a standard catalog glass is used.

The general data used in the design—the entrance pupil radius, the object height (OBY), the object distance (THO), which here is at infinity, and the image surface (IMS)—are given here. The paraxial constants are also given. In Figure 4.16 one can easily identify the effective focal length (EFL), f-number, and Petzval radius (PTZVAL), which can be seen to give the same results that were found in the earlier examples.

The example lens was generated in such a way as to minimize the primary and secondary lateral color (PLC and SLC) rather than the primary and secondary axial colors (PAC and SAC). The design program can be asked to test the earlier result. As can be seen, both the primary and secondary colors are minimal, of the order of 10^{-6}. The printout gives the values for each surface. One can see in Figure 4.17 that the interface of the two lenses has a chromatic aberration that the other two surfaces essentially cancel.

An important issue in the design of lenses is the edge thickness. For a given aperture it is important that each element of the system have a finite, positive thickness. Indeed, even positive lenses should have a reasonable thickness for mounting. The edge thick-

```
*CHROMATIC ABERRATIONS
 SRF       PAC           SAC             PLC            SLC
  1        --            --              --             --
  2    -0.001538     -0.001070       -0.000115     -8.02699E-05
  3     0.012887      0.009271      -5.02301E-05   -3.61371E-05
  4    -0.011483     -0.008118        0.000164       0.000116

 SM    -0.000688      0.000419      -7.62136E-06   -1.93444E-06
```

FIGURE 4.17.

```
*ETH FROM    2    TO    4 AT HTS    2.000000    &    2.000000    IS    0.955091
```

FIGURE 4.18.

nesses are generated as shown in Figure 4.18. The program can draw the lens based on the data input for the lens and one can see in Figure 4.19 that the flint element is meniscus with both radii having the same sign.

The next question that must be dealt with is the nature of the aberrations of the lens. This lens is to be used with an on-axis object at infinity. Thus the aberration of concern is spherical aberration. The OSLO program calculates both the third-order and the fifth-order spherical aberration as in Figure 4.20 (third-order) and Figure 4.21 (fifth-order) when the proper command is issued to the program. The values that are given by the program are coefficients of a polynomial that expresses the aberrations in terms of some ray parameters. Here one can see that the coefficient of the spherical aberration terms is significantly greater than any of the others. In the fifth-order aberrations the last column at the right contains the seventh-order spherical term, which can be seen to be significant in magnitude.

OSLO has graphics capability, as shown in the drawing of the lens. A rendering of the lens with bundles of rays entering the system from several angles is instructive. In Figure 4.22 three sets of rays have been drawn. The LSA that was seen to be significant in Example 4.2 is clearly visible in the ray trace of the rays which are parallel to the optic axis. The ray bundle at the greatest angle to the optic axis shows the coma present for off-axis rays with this lens if one looks closely at the crossings of the pairs of rays above and below the optic axis.

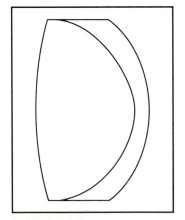

FIGURE 4.19.

*SEIDEL ABERRATIONS

SRF	SA3	CM3	AS3	PZ3	DS3	PA3
1	--	--	--	--	--	--
2	-0.001069	-8.01450E-05	-6.01087E-06	-9.15149E-06	-1.13718E-06	--
3	0.116946	-0.000456	1.77675E-06	7.40181E-06	-3.57762E-06	-8.77654E-11
4	-0.114162	0.001632	-2.33178E-05	-2.71901E-05	7.21843E-07	3.03594E-09
SM	0.004367	0.002788	-7.01164E-05	-7.36485E-05	-1.14802E-06	7.50276E-09

FIGURE 4.20.

*FIFTH-ORDER ABERRATIONS

SRF	SA5	CM5	AS5	PZ5	DS5	SA7
1	--	--	--	--	--	--
2	-1.10364E-05	-1.41098E-06	7.35538E-12	6.39231E-10	4.84940E-11	-1.33179E-07
3	0.057518	-0.000357	-4.84227E-11	-5.76017E-10	3.50390E-12	0.031275
4	-0.008992	0.000289	-5.35764E-10	2.30033E-09	-4.76335E-11	0.004039
SM	0.123465	-0.000176	-1.46797E-09	6.01494E-09	1.11067E-11	0.089872

FIGURE 4.21.

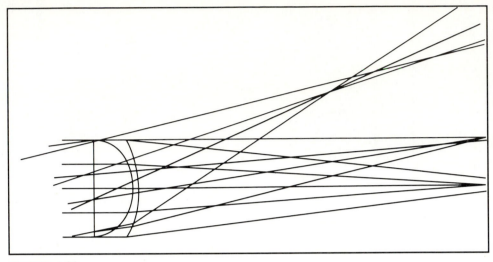

FIGURE 4.22.

There are a great number of additional things OSLO can do to aid in the design process. Aberrations can be plotted as a function of aperture, for example. One datum of interest to us is the spot diagram of Figure 4.23. This diagram gives the positions of rays traced through the system in the image plane. The symmetry of the spot diagram is to be expected since there is symmetry in this system about the optic axis. If one were examining a ray bundle from well off the optic axis, one would expect and indeed would find asymmetry.

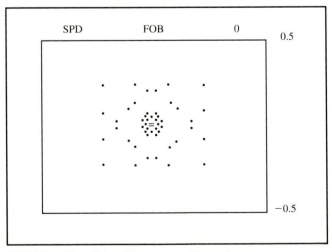

FIGURE 4.23.   The object is on the axis.

```
*YFAN
RAY             FY                 DYA          DXA              DY
 1         -1.000000           0.163925        --          -0.354412
 2         -0.800000           0.149875        --          -0.073431
 3         -0.600000           0.116315        --          -0.013763
 4         -0.400000           0.078334        --          -0.001707
 5         -0.200000           0.039275        --          -7.56237E-05
 6         -5.55112E-17        1.09064E-17     --          -1.27352E-32
 7          0.200000          -0.039275        --           7.56237E-05
 8          0.400000          -0.078334        --           0.001707
 9          0.600000          -0.116315        --           0.013763
10          0.800000          -0.149875        --           0.073431
11          1.000000          -0.163925        --           0.354412
```

FIGURE 4.24.

Precise ray tracing is, of course, part of the design process. Tracing a bundle of meridional rays from the axial point at infinity is instructive. FY in Figure 4.24 is the height of the ray on surface 1. All the rays are essentially parallel to the optic axis since they originate at infinity. DYA is the slope of the ray, and DY is its displacement relative to the optic axis. These data allow a check on the results of Example 4.2. The LSA is the ratio of DY to DYA and here we have

$$LSA = \frac{DY}{DYA} = \frac{0.354412}{0.163925} = 2.162038$$

where the change in sign of DYA is due to the different sign convention in the program. This value can be compared to the value gotten in Example 2 for the difference in position of the paraxial ray with the marginal ray

$$LSA = 12.062878 - 9.900820 = 2.162058$$

and one can see that the results are the same to within round-off error.

One of the important features of automated lens design programs is the automated correction provided. One can establish some design goals for a system, and the program will use correction algorithms to modify targeted parts of the system to bring the system into compliance with the design goals. The program thus can modify thicknesses, curvatures, and radii to produce a desired result. A note of caution should be injected here in that the program cannot do the impossible. A single element cannot be expected to be corrected for all the third-order aberrations or for color since there simply are not enough free variables in the two curvatures, one thickness, and one glass to fix a much larger number of aberrations; only one curvature is free since the other is needed to fix the power of the lens.

```
*OPERANDS INPUT
   OP      NAME          DEFINITION                    WEIGHT
   O 1                   "SA3+SA5"                     1.000000
   END

*VARIABLES
   VB   TYP SN CF            MIN        MAX        WGT
   V 1   CV  3  0            --         --        1.000000
   V 2   CV  4  0            --         --        1.000000
```

FIGURE 4.25.

The example lens was corrected for third- and fifth-order spherical aberration using the curvatures of the third and fourth surfaces. These quantities were entered into the program as shown in Figure 4.25.

The correction program required only a few iterations and was accomplished in less than one second. The ray trace from three angles, similar to that given in Figure 4.22 is shown in Figure 4.26.

This design requires a minimization of the on-axis image error. Comparing the on-axis set of rays with those in the earlier ray trace, one can see clearly that the lens has been significantly corrected for on-axis imaging. The oblique ray traces show that there has not been a significant change in the off-axis errors, but none is really expected. Little

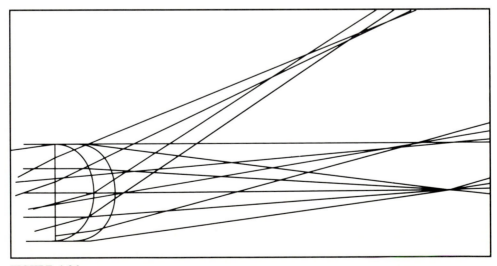

FIGURE 4.26.

```
*LENS DATA
Book Example
SRF     RADIUS            THICKNESS        APERTURE          GLASS  SPE   NOTES
1        --                  --            2.000000 A         AIR
2      15.000000         1.250000          2.000000 S          K5  C
3      -2.785602 V       0.200000          1.951013 S          F6  C
4      -5.999252 S       9.895471 S        1.953469 S         AIR
5        --                  --          S 0.101806 S         AIR

>pxc
*PARAXIAL CONSTANTS
          EFL            FNB              GIH              PIV           PTZRAD              TMAG
       10.180552      2.545138        0.101805        -0.020000      -14.026558        -1.018O6E-19
```

FIGURE 4.27.

change seems to have occurred in the shape of the lens as well, so it seems that some rather small changes in the curvatures have resulted in significant changes in the lens properties. The new lens parameter can be printed from the program, and Figure 4.27 shows that the changes are in fact quite small. The radius of the third surface has changed from −2.4700 to −2.7856 and the radius of the final surface from −5.7210 to −5.9993.

An examination of the aberration data from this new lens as shown in Figures 4.28 and 4.29 is interesting. Note that the aberrations have not gone to zero separately for the third-order spherical aberration and the fifth-order aberration; rather, the aberrations are equal and of opposite sign. One should note also that the seventh-order spherical aberration has not either gone to zero or been canceled.

The residual aberration can be tested by tracing a fan of meridional rays and evaluating the residual on-axis error for the marginal ray as in Figure 4.30. Proceeding, one finds

$$\text{Residual} = \frac{DY}{DYA} = \frac{0.079101}{0.188217} = 0.420265$$

a considerable reduction in the difference between the marginal and paraxial ray foci. This is but one simple example of an automated lens design as minimization targets.

This example has barely begun to examine the capability of automated design. What is important is that such programs now exist and the task of design of optical instruments can proceed at a much more rapid pace than was possible 50 years ago. The design task is, however, not one to be taken lightly. The designer is called upon to make many decisions in the course of the design process. Many of these require the designer to call upon previous experience since there is really no absolute design for a particular optical device, as there is no absolute design for an amplifier.

In subsequent chapters the testing of lenses will be discussed, as well as further design tools. The evaluation of optical systems depends largely on questions treated within the wave model.

```
>sei all
*SEIDEL ABERRATIONS
SRF      SA3           CM3            AS3            PZ3            DS3            PA3
1        --            --             --             --             --             --
2      -0.001069   -8.01450E-05   -6.01067E-06   -9.15149E-06   -1.13718E-06   -1.13320E-10
3       0.063818   -0.000408       1.98945E-06    6.56320E-06   -4.16677E-08    3.11420E-09
4      -0.105381    0.001569      -2.33732E-05   -2.59290E-05    7.34251E-07

SM     -0.057600    0.002751      -6.97230E-05   -7.25605E-05   -1.13155E-06    7.63766E-09
```

FIGURE 4.28.

```
>fif all
*FIFTH-ORDER ABERRATIONS
SRF      SA5            CM5            AS5            PZ5            DS5            SA7
1        --             --             --             --             --             --
2      -1.10364E-05   -1.41098E-06    7.35538E-12    6.39231E-10    4.84940E-11   -1.33179E-07
3       0.032994      -0.000256      -5.03965E-11   -5.27399E-10    4.07650E-12    0.014328
4      -0.010351       0.000310      -5.39209E-10    2.22250E-09   -4.82279E-11    0.001204

SM      0.057600       0.000134      -1.48191E-09    5.94120E-09    1.10525E-11    0.039531
```

FIGURE 4.29.

88

```
*YFAN
RAY              FY              DYA          DXA            DY
 1         -1.000000        0.188217        --       -0.079101
 2         -0.800000        0.156325        --       -0.001651
 3         -0.600000        0.118162        --       -0.006606
 4         -0.400000        0.078763        --       -0.003026
 5         -0.200000        0.039319        --        0.000442
 6      -5.55112E-17     1.09053E-17        --    -2.31172E-32
 7          0.200000       -0.039319        --       -0.000442
 8          0.400000       -0.078763        --       -0.003026
 9          0.600000       -0.118162        --       -0.006606
10          0.800000       -0.156325        --        0.001651
11          1.000000       -0.188217        --        0.079101
```

FIGURE 4.30.

PROBLEMS

Lens I:

The lens illustrated here is a double-gauss lens with a central stop. The lens is described by

Surface	1		2		3		4		5		6
Radius (cm)	62.		∞		101.20		−101.20		∞		−62.70
Thickness (cm)		1.11		0.74		2.78		0.74		1.11	
Index		1.744		1.649		1.00		1.649		1.744	

The stop is located at the center of the lens.

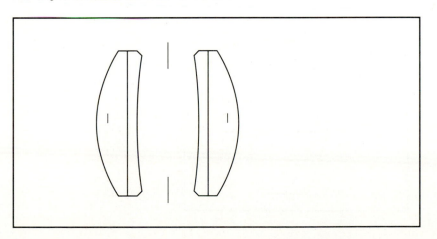

Lens II:

The following lens is an example of a Cooke triplet. The stop is located 0.02 cm before the fifth surface.

Surface	1		2		3		4		5		6
Radius (cm)	45.00		−500.0		−93.0		43.0		118.0		−63.0
Thickness (cm)		2.00		0.50		0.61		1.00		2.00	
Index		1.812		1.000		1.995		1.000		1.812	

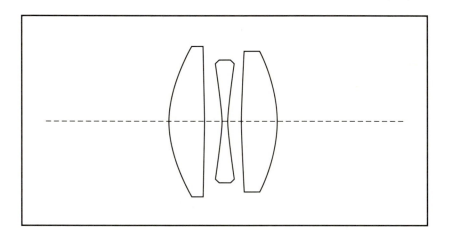

Lens III:

The following lens is an example of a simple doublet lens.

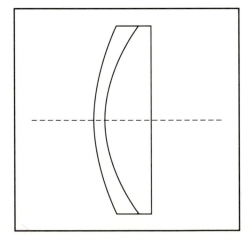

Surface	1		2		3
Radius (cm)	5.00		2.60		97.7
Thickness (cm)		0.27		0.75	
Index		1.720		1.617	
V		29.3		55.0	

The stop is located on the first surface.

4.1. Find the system matrix for lens I. Find the effective focal length, the back focal length, the positions of the principal planes, and the position of the entrance pupil and exit pupil.

4.2. Find the system matrix for lens II. Find the effective focal length, the back focal length, the positions of the principal planes, and the position of the entrance pupil and exit pupil.

4.3. Find the system matrix for lens III. Find the effective focal length, the back focal length, the positions of the principal planes, and the position of the entrance pupil and exit pupil.

4.4. Does lens III meet the criteria for achromaticity established in equation (4.18)?

4.5. Create an achromatic lens using K5 crown and F6 flint glasses where the third surface is to be plane.

4.6. Can one create an achromatic lens (system) with only one type of glass? If so, how?

4.7. Using the parameters of Example 2 and trigonometric methods, trace the marginal and zonal rays for an object distance of 20 cm. Is the lens better corrected for infinity or for 20 cm?

4.8. Using trigonometric traces, evaluate lens I for an axial object at infinity and for an axial object 200 cm before the lens. For which object is the lens better corrected?

4.9. Using trigonometric traces, evaluate lens II for an axial object at infinity and for an axial object 200 cm before the lens. For which object is the lens better corrected?

4.10. Using trigonometric traces, evaluate lens III for an axial object at infinity and for an axial object 20 cm before the lens. For which object is the lens better corrected?

The following problems are for those students with access to computerized design programs.

4.11. Examine lens I for objects at infinity and at 200 cm which subtend an angle of 20° at the first surface. For which condition is the lens best corrected?

4.12. Optimize lens I for an object at 200 cm subtending 20°.

4.13. Examine lens II for an object 1 m in front of the lens subtending 15°. Optimize the lens for coma and astigmatism, and compare the resulting lens with the original for objects at infinity.

4.14. Evaluate the color correction of lens III at 0.5 m and at infinity. How well optimized is the lens for color? What are possible changes one might make in the lens to improve its achromaticity?

CHAPTER 5

Optical Instruments

Until recently the design of optical instruments has been based almost entirely on the geometrical optical model. The ray tracing in the previous chapters has provided the detailed design data needed. Recently wave model concepts and calculations have found their way into the design process, but even today this process is largely geometrical. This chapter is devoted to the results of the design process, namely, to optical instruments. Only a few of the plethora of optical instruments which one encounters on an almost daily basis can be touched upon here, but the selection is meant to be general enough to give the reader a reasonably wide view of these instruments.

One fortunate aspect of optical design is that it has a modular aspect. Eyepieces are a common part of many direct viewing instruments, and these eyepieces can be considered a module that can be combined with various objective lenses to form a set of direct viewing optical devices, from microscopes to telescopes. In this chapter the modules that will be treated are eyepieces and objective lenses. These will be treated subsequently in combination as they are used in practical instruments. The module idea is one which can assist the engineer who needs to combine readily available off-the-shelf parts to aid in the creation of a prototype prior to the detailed design of a production device.

Eyepieces

An *eyepiece* is a device which acts on an object and presents its image to the eye in such a fashion that it subtends a much larger angle. Typically, the image is formed by other parts of the optical system, but this need not always be the case. Eyepieces, also known as *oculars*, are usually designed for the instrument with which they are being used, but, in principle, they are interchangeable. Since the image is presented at a large angle, the oblique aberrations are important, particularly astigmatism and field curvature. The chromatic aberrations also must be removed.

The simplest example of an eyepiece and one that is well suited to illustrate the principle of magnification is the simple magnifier. This magnifier is used directly with the object rather than with an intermediate object formed by another lens system. It is most easily illustrated by a simple biconvex lens often called a reading glass.

To view the detail in an object, typically one brings it closer to one's eyes. There is a limit to this, however, because eyes cannot accommodate to a very close object, so that at a certain distance the object no longer remains in sharp focus. Although this near point varies from individual to individual, the standard distance of 25.0 cm is taken as the point at which an object can be viewed in greatest detail. At 25.0 cm a given object will subtend an angle of ϕ. If a positive lens with the appropriate focal length is placed between the eye and the object, and the object is located inside the object-side focal point so that the image is presented at 25 cm, then a magnification of the object takes place. This is illustrated in Figure 5.1.

The system matrix for the simple magnifier is

$$\begin{bmatrix} 1 & -\dfrac{1}{f} \\ 0 & 1 \end{bmatrix} \tag{5.1}$$

and an object of height y which subtends an angle ϕ at 25 cm, that is, an angle $y/25$, will, when placed at a distance $-s_o$, subtend $y/(f - \delta)$ if s_o is $(f - \delta)$. The *angular magnification*, M, is then given by

$$M = \frac{y/s_o}{y/25} = \frac{-25}{s_o} = -\frac{25}{f - \delta} \tag{5.2}$$

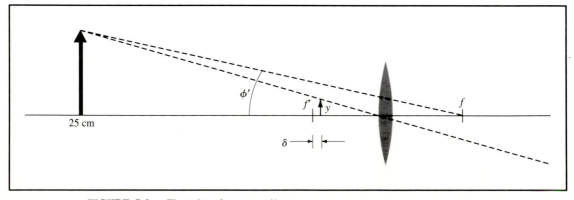

FIGURE 5.1. The simple magnifer.

The object distance $-s_o$ corresponds to an image distance of -25 cm for this system. Using equation (3.62)

$$s = \frac{-25}{1 + 25/f} \tag{5.3}$$

and

$$M = 1 + \frac{25}{f} \tag{5.4}$$

EXAMPLE 5.1   What is the angular magnification of an object viewed with a 20-D lens?

SOLUTION   Using equation (5.4) and noting that the focal length of a 20-D lens is $1/20 = 0.05$ m or 5 cm, one gets

$$M = 1 + \frac{25}{5} = 6\times$$

The notation used in the example for the dimensionless magnification M, namely, $6\times$, indicates that the angular magnification is 6 times. One must be careful to express the focal length in centimeters in equation (5.4) to satisfy the units.

Three other magnifiers are shown in Figure 5.2. The doublet (a) requires the object to be at the surface of the magnifier and is often used as a measuring instrument with a scale etched into the plano surface of the lens that rests on the object. The Coddington magnifier (b) made from a single piece of glass has a central stop created by the indentation about the center of the lens. Finally, (c) is an achromat, that is, a color-corrected version of the Coddington magnifier.

The Huygens eyepiece is shown in Figure 5.3. It consists of two lenses of the same glass set a distance t apart. The separation is fixed so that the system is color corrected. The power of the ocular is given by equation (3.42):

$$\phi = \phi_1 + \phi_2 - t\phi_1\phi_2$$

Differentiating, one gets

$$d\phi = d\phi_1 + d\phi_2 - t\phi_1\,d\phi_2 - t\phi_2\,d\phi_1$$

and using equation (4.12) with the one glass

$$d\phi = \frac{\phi_1}{V_1} + \frac{\phi_2}{V_1} - t\frac{\phi_1\phi_2}{V_1} - t\frac{\phi_1\phi_2}{V_1}$$

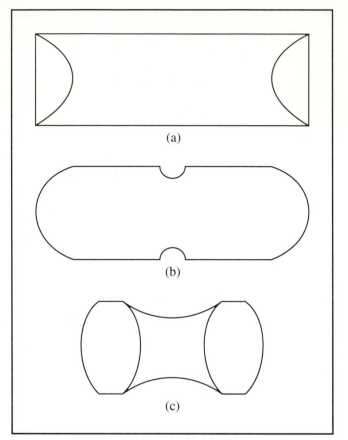

FIGURE 5.2. Three common magnifiers. (a) Doubler.
(b) Coddington. (c) Achromatized Coddington.

Setting $d\phi$ equal to 0 one gets

$$t = \frac{1}{2}\left(\frac{1}{\phi_1} + \frac{1}{\phi_2}\right) = \frac{1}{2}(f_1 + f_2) \tag{5.5}$$

and a spacing of the elements at a distance of one-half the sum of their focal lengths leads to a color-corrected eyepiece. At that spacing the overall power of the eyepiece is half the sum of the powers of the two lenses:

$$\phi = \frac{1}{2}(\phi_1 + \phi_2) \tag{5.6}$$

The first lens in the Huygens eyepiece is called the *field lens* and the second is the *eye lens*. The field lens commonly has a focal length of 1.5 times that of the eye lens. The

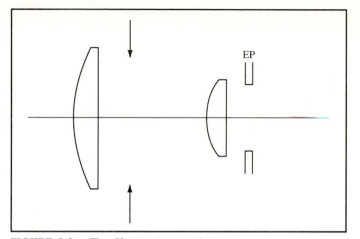

FIGURE 5.3. The Huygens eyepiece.

field lens forms a real image of the image from the objective of the system at the position of the arrow in Figure 5.3. As a result, this eyepiece cannot serve as a magnifier. At the image position of the field lens one places a field stop. The exit pupil, EP in the figure, corresponding to an entrance pupil formed by the objective of the system, is positioned after the eyepiece at the position shown in the figure. This allows the pupil of the eye to come close to the exit pupil of the system, although the *eye relief* of the system, the distance from the eyepiece lens to the exit pupil, is generally shorter than the most comfortable distance.

While the system is corrected for primary lateral chromatic aberration, the individual lenses are not, so that if one puts cross-hairs or a recticle at the field stop, it will show distortion and color. The Huygens eyepiece, even though color corrected, has astigmatism, rather large pincushion distortion, and a small Petzval radius. Today it is used principally with inexpensive, low-power microscopes and telescopes.

The Ramsden eyepiece, Figure 5.4, is similar to the Huygens eyepiece but it is made with two lenses of the same glass with equal focal lengths. The separation of the elements is then the focal length of one of them. The object-side focus corresponds to the position of the field lens so that cross-hairs or a reticle must be placed there. A problem arises in this regard in that dust on the field lens appears in the image, and to alleviate this problem the lenses are placed at a separation slightly less than f so that the focal plane lies slightly ahead of the system.

Reducing the separation introduces some lateral color, but the Ramsden eyepiece has no coma and considerably less distortion. The eye relief is also greater, making this a more comfortable eyepiece to work with, particularly if it is to be used over long periods of time. The lateral color that was reintroduced with the reduced spacing can be corrected by using a cemented doublet as the eye lens to provide color correction. With this configuration the eyepiece is known as the Kellner eyepiece and is the eyepiece of choice where cross-hairs are required, as well as in prism binoculars.

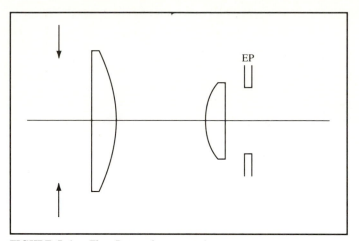

FIGURE 5.4.   The Ramsden eyepiece.

The field of the eyepieces just discussed is generally less that 45°. Various wide-angle eyepieces have been developed often for use in gun sights. One example, the Erfle, is shown in Figure 5.5. It has a large eye relief, an obvious necessity with recoiling guns, and a field 1.5 times that of the equivalent Kellner eyepiece.

Objectives

An *objective lens* forms the initial image in an optical system. It is called an objective lens since it is the lens closest to the object. The image formed by the objective lens may be an intermediate image as in a telescope or microscope, or it may be the image that is used by the system, as with photographic lenses.

A photographic system requires a sharp, bright image of the scene being photographed. Ideally, the best camera would be of the pinhole type in which the entire field is in sharp focus and aberrations are absent; however, far too little light energy passes through the pinhole to expose even the most sensitive of films. It can be shown that the density of light entering a camera is directly proportional to the area of the entrance pupil and inversely proportional to the square of the focal length. Since the $f/$ number is defined as f/a, where a is the diameter of the aperture, the light energy falling on the film is proportional to $1/(f/)^2$. The smaller the $f/$, the ''faster'' the lens. Modern cameras have an adjustable calibrated pupil given in terms of $f/$, most often in the sequence 1, 1.4, 2.0, 2.8, 4.0, 5.6, 8.0, 11.0, 16.0, 22.0, which represents an approximate halving of the light energy falling on a film per unit time for each stop change.

The simplest and cheapest camera objective is the landscape lens, a simple meniscus lens used as shown in Figure 5.6. This was the lens used in the original box Brownie,

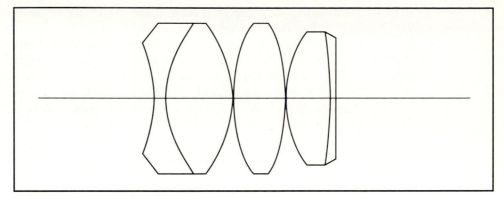

FIGURE 5.5. The Erfle eyepiece. The eye relief is about 8 times that of the equivalent Huygens eyepiece.

the first mass-production camera from Kodak. This form of lens is used in spite of a large spherical aberration because it possesses a relatively flat field. The effective field is less than 40° because of other off-axis aberrations. Since the Brownie was originally used with black-and-white film, it was not necessary to color correct this lens. When color film is used with this lens, the lateral color problem is obvious.

The problems with the simple meniscus have been attacked in a number of ways. Initially, it was achromatized, but the peripheral field remained a problem, particularly the astigmatism. The solution was to go to a lens that was symmetrical about a center stop. Such symmetrical lenses, when used between equally spaced conjugates, have coma, distortion, and lateral color corrected as a result of the symmetry. When the conjugates are not equally spaced, that is, when the object distance is not equal to the image distance as

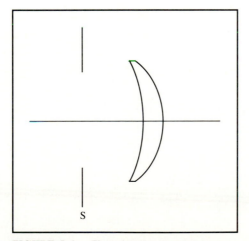

FIGURE 5.6. The simple landscape lens showing the stop S which lies on the object side of the lens.

in the case of a camera, the correction is not total and additional corrections need be made. The symmetrical lens does provide a good starting point for the design, however.

Figure 5.7 shows three forms of symmetric lenses. The Rapid Rectilinear (a) functions well only at small aperture, usually $f/11$ or smaller, since the Petzval curvature is significant. The Zeiss Topogon (b) was not only an improvement on the Rapid Rectilinear, but it has a half-field of the order of 45°. Most symmetric lenses used today are used to take advantage of the wide field. One of the most popular wide-angle lenses is the Double Gauss (c), which is widely used today in some modification.

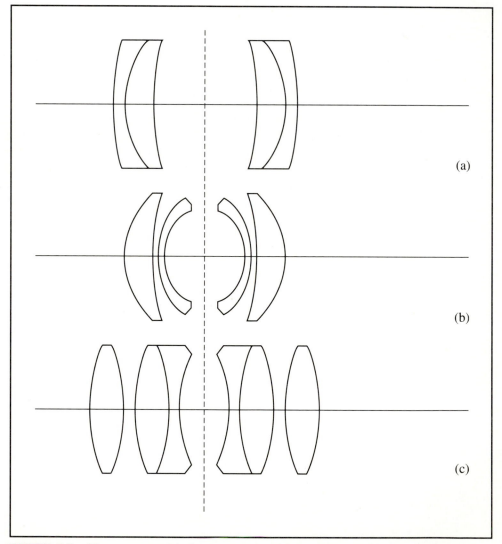

FIGURE 5.7. Three symmetric lenses. (a) Rapid Rectilinear. (b) Zeiss Topocon. (c) Double Gauss.

A second important class of camera objective lenses is based on the Cooke triplet designed by H. Dennis Taylor and manufactured by the Cooke Company in England, from which it derives its name. Taylor designed the triplet to solve the problem of field curvature. The result shown in Figure 5.8a has a smaller half-field, only about 20°, than the symmetric lenses but with relatively fast apertures of $f/3.5$. This fast aperture contributed importantly to the development of motion pictures.

As with the symmetric lens, many design modifications were made. Literally hundreds of lenses derived from this basic design and that of the Zeiss Tessar, Figure 5.8b, which was produced shortly after the Cooke triplet. One example of a triplet derived from

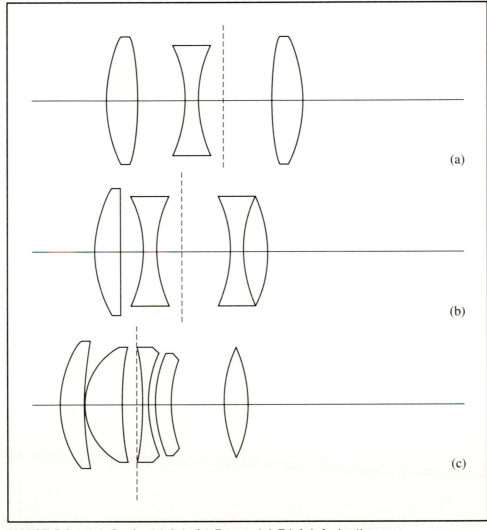

FIGURE 5.8. (a) Cooke triplet. (b) Tessar. (c) Triplet derivative.

the earlier prototypes is shown in Figure 5.8c. The elements of the original triplet have been broken up, giving more variables to be used to reduce aberrations.

Other classes of objective lenses used in cameras include telephoto lenses and zoom lenses. The former are in simplest terms telescopes with a positive power, while the latter are designed so that some elements of the lens system can be moved, changing the image magnification without changing the position of the image plane. Simple examples of these lenses are shown in Figure 5.9.

The telephoto lens is simply a telescope with a finite back focal plane at the film plane. Other forms of telephoto include an afocal system which may be attached directly to the lens system of a camera to provide telephoto capability.

The zoom lens in Figure 5.9b is a mechanically compensated zoom lens. The system consists of a telescope of variable magnification as the first lens pair in the figure coupled to a primary lens which provides an image at the film plane. The lens is mechanically compensated since the first two lenses not only move but move relative to each other, the first positive lens making a forward and then a regressive motion while the second lens moves only toward the primary lens. In optically compensated zoom lenses the moving elements move as a unit with no relative motion. Typically, if m is the maximum magnification, the minimum is $1/m$, and the focal length variation of the lens is of the order of m^2. Zoom lenses are widely used today in home video camera systems, which typically have a magnification range of 6 to 8 times.

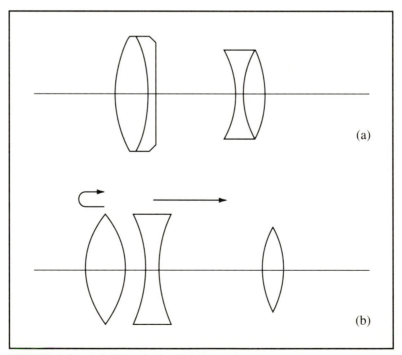

FIGURE 5.9.   (a) Telephoto. (b) Zoom lens.

Telescope objectives require a very good image on the axis. Generally, these instruments need a large aperture to gather as much light as possible and because of their manner of use require that the LSA and the longitudinal chromatic aberration be corrected. Often in mass-produced objectives for binoculars, for example, the field aberrations are largely ignored or are only partially corrected.

The simplest telescopic objective is the doublet, which as was seen in the design example in Chapter 4, can be corrected for both LSA and chromatic aberration. The doublet need not be a cemented or contacted doublet since the separation of the elements gives an additional element for aberration control. Occasionally the positive element is split into two elements to control the secondary chromatic aberration, as in Figure 5.10.

Microscope objectives are designed to magnify fine detail in an object to make the normally invisible visible. The overall magnification of the microscope occurs in two steps, with the primary magnification coming from the objective and the secondary magnification from the eyepiece. Objectives normally range from $10\times$ to $200\times$, with the magnifying power typically engraved on the barrel of the objective.

The microscope objective is designed to provide a real image at a fixed reference position in the microscope tube, generally using a standard optical tube length of 160 mm and to give this image when at a distance d from the object. This then permits the interchange of eyepieces without extensive alteration of the optical system; alternate eyepieces can simply be dropped into the tube. The distance d is the microscope working distance, varying from several millimeters for low-powered objectives to several tens of microns (μm) for the high powered. There is some variability in d for which no change in image sharpness is observed and this range is called the *depth of focus*. The depth of focus is somewhat wider for visual instruments than for photographic ones and ranges from 10 μm or so for low-powered objectives to a fraction of a micron for the high-powered objectives.

Since the microscope is basically an on-axis system, the designer is concerned with spherical aberration, chromatic aberration, and field curvature. The other aberrations are generally of little consequence in properly designed systems.

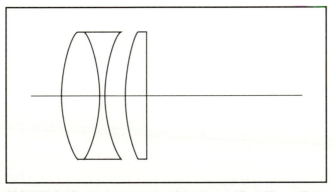

FIGURE 5.10. A telescopic objective with split positive element.

All useful microscope objectives are achromatized, and a majority of these objectives are simple achromats as in Figure 5.11a. Secondary color, small chromatic aberrations existing after chromatic aberration is corrected by the technique presented earlier, is also corrected in lenses known as apochromats, Figure 5.11b, which use fluorite mineral crystal lenses in combination with glass lenses. Fluorite is used because of its low index and large dispersion. Apochromats cost several times more than the equivalent achromat, and some "semiapochromats" are available which use less fluorite and contain fewer elements and are correspondingly less expensive than the apochromats.

From Figure 5.11 one can see that microscope objectives have a number of steep curves and therefore a short Petzval radius. With the short Petzval radius, the field curvature becomes a significant problem. There are a number of objective designs which compensate for this problem and provide some field flattening, but these designs tend to be very complex, with as many as 15 elements.

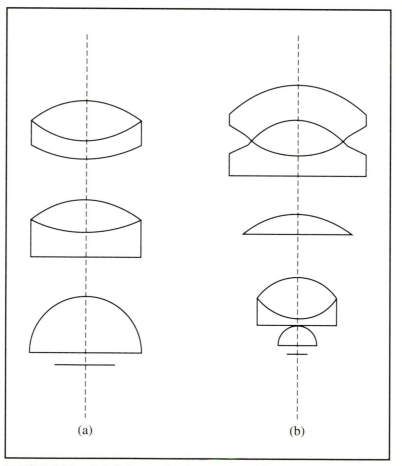

(a)   (b)

FIGURE 5.11.   (a) Achromat microscope objective.
(b) Apochromat microscope objective.

Compound Visual Instruments

The combination of an eyepiece with an objective results in a compound visual instrument. There are two principal functions of these instruments, both involving magnification: the magnification of distant objects with telescopes and the magnification of small, near objects with microscopes.

Microscopes are direct view systems, as shown in Figure 5.12. The objective forms an intermediate, highly magnified, real image at position I in the microscope tube. The intermediate object I falls just inside the focal position of the eyepiece, which has its virtual image magnified and presented at I', which is the image viewed by the observer's eye. Note that I is inverted relative to the object O. Since I' is erect relative to I, the final image is inverted. If the final image I' is at 25 cm from the observer's eye, the overall magnification is given by

$$M = m_o m_e \qquad (5.7)$$

where, if w is the working distance and f_o the focal length of the objective,

$$m_o = -\frac{w}{f_o}$$

and

$$m_e = \frac{25}{f_e}$$

with f_e the eyepiece focal length. The total magnification is then

$$M = -\frac{25w}{f_o f_e}$$

Usually m_o and m_e will be labeled on the barrel of the element.

Telescopes are somewhat more varied than microscopes. There are two principal types: the astronomical telescope, in which the image is inverted, and the terrestrial telescope, with an erect image. The power of the telescope is the ratio of the angle subtended at the eye by the image to the angle subtended by the unmagnified object. It is necessary here to use angles since both the object and the image are effectively at an infinite distance.

Figure 5.13 illustrates the simple astronomical telescope. The objective forms an inverted image of the infinitely remote object at an intermediate position I'. The eyepiece is set so that I' is at or near the object-side focal plane. The resulting image is inverted relative to the original object. The image may be back at infinity if the intermediate object

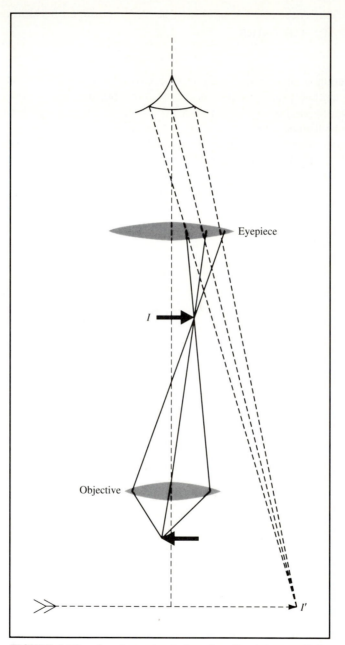

FIGURE 5.12.   A microscope showing the intermediate
image I and the final image I'.

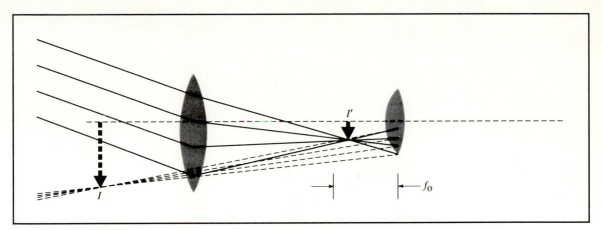

FIGURE 5.13. The astronomical telescope.

is at the eyepiece focal plane, or it may be displayed at an intermediate distance such as 25 cm, as shown in the figure. The magnification is the ratio

$$M = \frac{\theta'}{\theta} \tag{5.8}$$

since with infinitely distant objects, it is necessary to use angular magnification. This ratio also may be expressed in terms of the focal lengths

$$M = -\frac{f_o}{f_e} \tag{5.9}$$

where f_o is the objective's focal length and f_e is that of the eyepiece. The length of the telescope, L, becomes the sum of the focal lengths,

$$L = f_e + f_o \tag{5.10}$$

The inverted image is at best distracting when the astronomical telescope is used to view distant objects on the earth, where a vertical reference is always present. The earliest telescopes, such as that used by Galileo, overcame this problem with a negative eyepiece, as shown in Figure 5.14. Not only is the image erect relative to the distant object, but the length of the telescope becomes $f_o + (-f_e)$ and is much more easily handled since it is shorter.

A second approach to providing an erect image in a telescope involves the use of a relay lens operating at unit magnification between the image formed by the objective and the eyepiece as in Figure 5.15. The relay lens inverts the image formed by the objective, and this leads to a final erect image. It is easily shown that the length of the telescope is increased by four times the focal length of the relay lens. One can recall the sea captain's

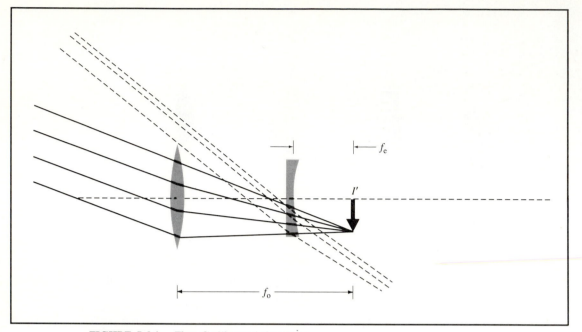

FIGURE 5.14. The Galilean telescope.

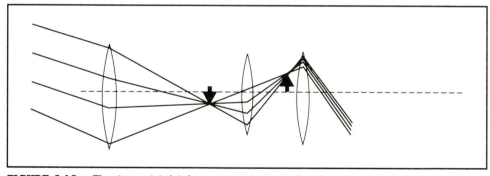

FIGURE 5.15. The terrestrial telescope using a relay lens.

telescope in movies about sailing ships and realize the difficulty associated with handling the long tube of the telescope.

Binoculars are simply telescopes mounted side by side. An interesting approach to the image inversion is taken with prism binoculars, where the use of a pair of prisms set at right angles to each other and acting as reflectors provides a correction to the inverted image. The first prism reinverts the image and the second flips right and left. In addition, the prisms increase the optical path length between the objective and eyepiece, allowing for greater magnification. Generally the magnification of binoculars lies in the $5\times$ to $10\times$ range. The upper limit is due to the fact that higher magnification would make the size and weight such that a rigid mount would be required for their use.

The field of view in telescopes and binoculars is fixed by the eyepiece aperture, and generally this is made as large as practicable. The diameter of the objective fixes the amount of light energy in the image. This is particularly important with night glasses, which are used in low-light environments. The specifications of binoculars are given as 7×35, for example, which denotes a 7-power telescope with a 35-mm-diameter objective. The exit pupil diameter becomes 35/7 or 5 mm if the entrance pupil is the objective, as is typically the case. The 5-mm exit pupil is sufficient for daylight use, but for night use one would usually use a 7×50 telescope with a larger exit pupil to accommodate the dark-adapted, dilated pupil of the observer.

One important use of telescopes is as collimating devices. A *collimator* is a device which provides a beam of parallel light for use in testing devices and in alignment of optical systems. The simplest collimator is simply a pinhole source set at the focal point of a converging lens. Pinhole sources often have too little intensity to be useful. Telescopes have the ability to image objects at infinity and thereby function as collimators. The beam from such a device is parallel, but if the source is extended, the beam will be divergent. In practice, trade-offs are made between the size and distance of the source and the divergence of the output beam of the collimator.

Spectacles

The final optical device to be presented here is the most common optical device is use today, the spectacle lens. Roughly 50% of young adults benefit from spectacle or contact lens correction. In the population group beyond age 50 nearly everyone requires some sort of spectacle.

The optical schematic of the human eye is shown in Figure 5.16. The eye is a spherically shaped organ about 24 mm long with a transparent spherical projection, the cornea, at its most forward point. The total power of the eye is about 60 D, and the image is formed on the retina. There are two powered optical elements in the eye: the cornea with a power of about 45 D and the crystalline lens. The crystalline lens is a unique optical element in that it can change its shape and thus allow the eye to change its object distance from infinity to about 10 cm for a 25-year-old while maintaining an image on the retina. Often, however, as a result of an improper eye length or a cornea which is not appropriately shaped for the size of the eye, the image on the retina is defocused and not sharply formed. If the power of the cornea and the lens is too small, the image of a distant object falls behind the retina, and the person is said to be hyperopic or farsighted. A converging spectacle can be provided to restore the image plane to the retinal plane. Similarly, if the eye is too long or the cornea too steep, the image is formed in front of the retina, and a negative, diverging lens is necessary to bring the image plane to the retina.

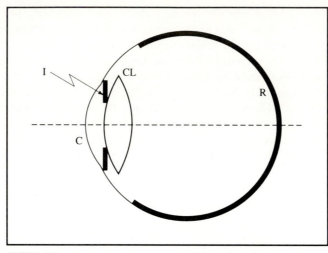

FIGURE 5.16.   The human eye. C is the cornea, I the iris, CL the crystalline lens, and R the retina.

Such a condition is known as myopia or nearsightedness. Once the distance correction is made with a corrective spectacle or contact lens, the natural focal powers of the crystalline lens provide near and intermediate focus.

As one ages, the crystalline lens loses its ability to alter its power and the person is said to be presbyopic (old-eyed). This means that to view near objects an additional correction is required beyond that needed for distance. Spectacles with two focal regions are called bifocals. Typically, when no more accommodation remains, generally at about age 60, the near segment of the bifocal is 3 diopters stronger than the distance portion. This gives a near point of about 33 cm, a suitable reading distance. Individuals with special needs at an intermediate distance, such as musicians, whose music may be a meter or more from the eye, often have trifocals, with the intermediate power set in a region between the distance and near portions.

PROBLEMS

5.1. A magnifier, type (a) in Figure 5.2, is made of two planoconvex lenses with focal length 4 cm set 3.8 cm apart. Why are they not set at 4 cm? What is the magnifying power of the system, assuming image formation at 25 cm?

5.2. A Huygens eyepiece is made with two thin lenses with powers 40 D and 66.7 D set so as to eliminate chromatic aberration. Find the focal length and magnification.

5.3. Where would one place cross-hairs in a Huygens eyepiece?

5.4. A magnifier is made in the form of a Ramsden eyepiece as in Figure 5.4. If each of the lenses has a power of 40 D, what magnification will occur for an object on the planar surface of the field lens?

5.5. If the separation of the elements in Problem 5.4 is reduced to 1.5 cm, what will be the magnification of an object at the surface? 2 cm before the surface?

5.6. The lens on my camera is labeled 50 mm $f/1.7$. (a) What is the maximum aperture of the lens? (b) If the diameter of the aperture is reduced to one-half its maximum value, how much longer would one have to expose the film compared with the full aperture?

5.7. The *depth of field* of a camera lens is the incremental distance over which objects will be "in focus" for some criterion. Consider an $f/3.5$ lens with a 55-mm focal length and the criterion that the maximum diameter of the circle of least confusion is 40 μm. An object is 200 cm in front of the camera. What is the depth of field?

5.8. The *depth of focus* is the object region that meets a criterion such as that specified in Problem 5.7. What is the depth of focus for Problem 5.7?

5.9. The original microscope was simply a glass sphere which rested on the object to be magnified. If the sphere has an index of 1.612 and a diameter of precisely 2.0 cm, what will be the image position and magnification?

5.10. A microscope has an eyepiece with focal length 10 mm and an objective with focal length 2.5 mm. The objective forms its image 160 mm beyond its image-side focal plane. What is the total magnification?

5.11. The eyepiece and objective of a microscope have focal lengths of 6.0 and 8.0 mm, respectively. They are 180 mm apart. Treat the microscope as thin lenses with the image -250 mm from the eyepiece lens. Where is the object located relative to the objective lens? What is the linear magnification produced by the objective? What is the overall magnification of the microscope?

5.12. The eye relief of a viewing system is the distance from the last lens of the system to the exit pupil. An astronomical telescope has an objective 2.0 cm in diameter with a focal length of 9.5 cm. The eyepiece used with it has a focal length of 2 cm and is 1.5 cm in diameter. What is the eye relief? What is the angular magnification? What is the diameter of the exit pupil?

5.13. A terrestrial telescope is to be 30 cm long and uses a 10-D objective lens and an 30-D eyepiece lens. Describe the relay lens and find where it should be placed in the tube. What is the power of the telescope?

5.14. A telephoto camera lens in one form consists of a Galilean telescope attached to the lens of a camera. Describe the system. Where should the camera lens be focused prior to attaching the telescope? What is the effect on the depth of field?

5.15. A pair of binoculars has objective lenses 7.5 cm in diameter with 28-cm focal length and an eyepiece with a -2.5-cm focal length whose aperture is 12.5 mm. What is the field of view at 1000 m?

5.16. The near point of the human eye is 250 mm, and the length of the eye is 25 mm. Find a reasonable approximation to the power of the eye focused at infinity. What additional power is required to accommodate to the near point? Assuming that a person has a far point of 1 m, what spectacle lenses would correct her distance vision and what would be the effect on her 25-mm near point?

CHAPTER 6

Optical Waves

Ray tracing and geometrical optics derive essentially from a linear propagation model for light. This linear propagation model is, as has been seen in the previous chapters, extremely useful for the understanding of the imaging properties of lenses and mirrors and for layout and design of optical instruments. As useful as it may be, it is unable to cope with certain observations, namely, interference and diffraction. To understand these observations, one must use a different computational model, the wave model. This chapter will serve as an introduction to this model.

Maxwell's Equations

The axiomatic approach to the wave model of light begins with Maxwell's equations, which we can state in their differential form as

$$\nabla \cdot \mathbf{E} = \frac{\rho}{\epsilon} \tag{6.1a}$$

$$\nabla \times \mathbf{E} = -\mu \frac{\partial \mathbf{H}}{\partial t} \tag{6.1b}$$

$$\nabla \cdot \mathbf{H} = 0 \tag{6.1c}$$

$$\nabla \times \mathbf{H} = \mathbf{j} + \epsilon \frac{\partial \mathbf{E}}{\partial t} \tag{6.1d}$$

where $\mathbf{E}$ and $\mathbf{H}$ are the electric and magnetic field vectors, respectively, ρ is the charge density, and $\mathbf{j}$ is the current density. The quantity ϵ is the permittivity of the medium in which the propagation is taking place, and μ is the permeability of the medium. In optics one deals primarily with media which are free of sources of charge and current where Maxwell's equations take the form

$$\nabla \cdot \mathbf{E} = 0 \tag{6.2a}$$

$$\nabla \times \mathbf{E} = -\mu \frac{\partial \mathbf{H}}{\partial t} \tag{6.2b}$$

$$\nabla \cdot \mathbf{H} = 0 \tag{6.2c}$$

$$\nabla \times \mathbf{H} = \epsilon \frac{\partial \mathbf{E}}{\partial t} \tag{6.2d}$$

in such regions.

If one takes the curl of both sides of equation (6.2b), one gets

$$\nabla \times \nabla \times \mathbf{E} = -\mu \nabla \times \frac{\partial \mathbf{H}}{\partial t} \tag{6.3}$$

Using the vector identity

$$\nabla \times \nabla \times \boldsymbol{\zeta} = \nabla(\nabla \cdot \boldsymbol{\zeta}) - \nabla^2 \boldsymbol{\zeta}$$

and the fact that ∇ and $\partial/\partial t$ commute, one gets

$$\nabla(\nabla \cdot \mathbf{E}) - \nabla^2 \mathbf{E} = -\mu \frac{\partial}{\partial t}(\nabla \times \mathbf{H})$$

which, following substitution from equations (6.2a) and (6.2d), yields

$$\nabla^2 \mathbf{E} = \mu\epsilon \frac{\partial^2 \mathbf{E}}{\partial t^2} \tag{6.4}$$

A similar wave equation is also gotten for $\mathbf{H}$.

Comparing this to the standard form of the wave equation, one can identify the propagation velocity of the wave with the quantity $(\mu\epsilon)^{-1/2}$.

$$\nabla^2 \psi = \frac{1}{v^2} \frac{\partial^2 \psi}{\partial t^2} \tag{6.5}$$

If the propagation is in free space so that

$$\epsilon = \epsilon_0 = \frac{1}{36\pi} \times 10^{-9} \ \text{F/m}$$

and $\mu = \mu_0 = 4\pi \times 10^{-7}$ H/m, one sees that $v = c = 3 \times 10^8$ m/s, the well-established velocity of light. One of the important results of Maxwell's theory and equations was just this, the demonstration that electromagnetic waves propagate at the velocity of light. One then links light waves with electromagnetic waves.

When the propagation takes place in a material medium, $\epsilon = \epsilon_r \epsilon_0$ and $\mu = \mu_r \mu_0$, where ϵ_r and μ_r are the relative permittivity and permeability, respectively. Here the velocity is $c/(\epsilon_r \mu_r)^{1/2}$, where c is the free-space (vacuum) velocity of light. If one recalls that the index of refraction n of a medium is given by the ratio c/v, then $n = (\epsilon_r \mu_r)^{1/2}$. In most optical media $\mu_r \approx 1$, and the refractive index can be approximated by $\sqrt{(\epsilon_r)}$. The relative permittivity of benzene is 2.284 and its index of refraction is 1.5132 for the F-line (486.13 nm). Using the approximation here, one would predict that $n = 1.5116$.

Solutions to the Wave Equation

Once the wave equation (6.4) has been established, one needs to find its solutions. There are three direct, closed-form solutions that are of importance here, and more complex solutions may be expressed as a combination of these solutions. Equation (6.4), when written in rectangular coordinates for a vacuum,

$$\frac{\partial^2 \mathbf{E}}{\partial x^2} + \frac{\partial^2 \mathbf{E}}{\partial y^2} + \frac{\partial^2 \mathbf{E}}{\partial z^2} = \frac{1}{c^2} \frac{\partial^2 \mathbf{E}}{\partial t^2} \tag{6.6}$$

can be seen to contain a spatial part, the left-hand side, and a time-dependent part, the right-hand side. If the propagation is taking place in a medium, c can be replaced by v in that medium. If one writes $\mathbf{E}$ in the form $\mathbf{E} = \mathbf{E}_s(x,y,z)T(t)$, that is, if one assumes that the spatial and the time parts of $\mathbf{E}$ are separable functions, then substituting this in (6.6) and dividing each side by the $\mathbf{E}$ one gets

$$\frac{1}{\mathbf{E}_s(x,y,z)} \left(\frac{\partial^2 \mathbf{E}_s}{\partial x^2} + \frac{\partial^2 \mathbf{E}_s}{\partial y^2} + \frac{\partial^2 \mathbf{E}_s}{\partial z^2} \right) = \frac{1}{c^2 T(t)} \frac{\partial^2 T^2}{\partial t^2} \tag{6.7}$$

At a fixed spatial position, as time varies the left-hand side remains fixed, while for a fixed time, the spatial function $\mathbf{E}_s$ varies while the right-hand side remains fixed. This condition can be satisfied if each side is equal to the same constant, which will be taken as $-q^2$. Thus for the time-dependent part

$$\frac{1}{c^2 T(t)} \frac{\partial^2 T(t)}{\partial t^2} = -q^2 \tag{6.8a}$$

or

$$\frac{d^2T(t)}{dt^2} + q^2c^2T(t) = 0 \tag{6.8b}$$

where the equation is now written in ordinary form since only one variable appears in it. The solution of (6.8b) is

$$T(t) = c_1e^{jqct} + c_2e^{-jcqt} \tag{6.9}$$

where c_1 and c_2 are constants depending on the initial conditions. The quantity qct must be dimensionless as the argument of a transcendental function. The quantity qc, which has dimension t^{-1}, is the angular frequency ω of the wave, and (6.9) can be written

$$T(t) = c_1e^{j\omega t} + c_2e^{-j\omega t} \tag{6.10}$$

which will be used here in the equivalent form

$$T(t) = c_0e^{-j(\omega t + \phi_1)} \tag{6.11}$$

The constant q is $2\pi/\lambda$, where λ is the wavelength in the medium in which the propagation is occurring.

The solution of the spatial part depends on the form of ∇^2 in equation (6.5). If rectangular coordinates are used as in (6.6),

$$\frac{\partial^2\mathbf{E}_s}{\partial x^2} + \frac{\partial^2\mathbf{E}_s}{\partial y^2} + \frac{\partial^2\mathbf{E}_s}{\partial z^2} = -\frac{4\pi^2}{\lambda^2}\mathbf{E}_s \tag{6.12}$$

then

$$\mathbf{E}_s = \mathbf{E}_{os}\exp j\left[\left(\frac{2\pi x}{\lambda_x} + \frac{2\pi y}{\lambda_y} + \frac{2\pi z}{\lambda_z}\right) + \phi_2\right] \tag{6.13}$$

where $\mathbf{E}_{os}$ is the wave amplitude and ϕ_2 is a phase to satisfy the need for two constants of integration. The quantity $2\pi/\lambda$ that appears frequently in wave solutions is given a special notation $\mathbf{k}$ and is called the *wave vector:*

$$\mathbf{k} = \frac{2\pi}{\lambda_x}\hat{\mathbf{i}} + \frac{2\pi}{\lambda_y}\hat{\mathbf{j}} + \frac{2\pi}{\lambda_z}\hat{\mathbf{k}} \tag{6.14}$$

Its direction is the direction of propagation of the wave. Using this, equation (6.12) can be written

$$\mathbf{E}_s(x,y,z) = \mathbf{E}_{os}e^{j(\mathbf{k}\cdot\mathbf{r} + \phi_2)} \tag{6.15}$$

and the solution of (6.7)

$$\mathbf{E}(x,y,z,t) = \mathbf{E}_oe^{j(\mathbf{k}\cdot\mathbf{r} - \omega t + \phi_0)} + \mathbf{E}_oe^{j(\mathbf{k}\cdot\mathbf{r} + \omega t + \phi_0)} \tag{6.16}$$

where $\phi_0 = \phi_1 + \phi_2$ and $E_o = E_{os}c_0$.

The vector solution of equation (6.12) contains the vectors $\mathbf{k}$ and $\mathbf{E}_o$. The wave vector $\mathbf{k}$ appears as an inner product with the position vector $\mathbf{r}$ so that the vector solution has its directions established by $\mathbf{E}_o$, the amplitude. Electromagnetic waves are *transverse waves;* that is, the displacement associated with the wave is normal to the direction of propagation of the wave. As a result, the inner products $\mathbf{E}_o \cdot \mathbf{k}$ will always vanish.

The waves in this solution are *plane waves. Plane waves* are characterized by the condition that the amplitude is independent of the position so that $\mathbf{E}_o$ does not depend on $\mathbf{r}$ but is rather a constant vector, and the constant phase surfaces are planes. The direction of $\mathbf{E}_o$ is called the *polarization direction* of the wave.

The first term on the right-hand side of (6.16) represents the forward-going wave while the second term represents a wave traveling in the negative direction. A simple rule to remember is that if the signs of the spatial term and time term in the exponent are different, then the wave is traveling in the positive direction. The forward-traveling plane wave is given by

$$E(x,y,z,t) = \mathbf{E}_o e^{j(\mathbf{k}\cdot\mathbf{r} - \omega t + \phi_0)} \qquad (6.17)$$

Figure 6.1 shows equiphase planes for a plane wave propagating in the direction $\mathbf{k}$. The lines represent constant phase planes separated by π. The origin is taken at a zero-phase position. The phase at P is given by $2\pi/\lambda$ times the projection of $\mathbf{r}$ onto $\mathbf{k}$. The apparent wavelength along the x axis can be seen to be $\lambda/\cos\theta$ and along the y axis to be $\lambda/\sin\theta$.

In general, only one component of $\mathbf{E}_o$ will be taken into account, in which case the solution will be taken in a scalar form and the solution (6.16) will be written

$$E(x,y,z,t) = E_o e^{-j(\omega t - \mathbf{k}\cdot\mathbf{r} + \phi_0)} \qquad (6.18)$$

In those cases involving polarization, the vector form will be reintroduced.

EXAMPLE 6.1

Given a wave

$$\mathbf{E}(x,y,z,t) = (\sqrt{2}\hat{\mathbf{i}} + 3\hat{\mathbf{j}} + \sqrt{5}\hat{\mathbf{k}})e^{-j(1.5\times10^8)t - 3x + \sqrt{2}y)}$$

Find the direction of propagation, the polarization direction, the magnitude of the amplitude, the wavelength, the frequency, and the velocity of propagation.

Solution

The direction of propagation is $(3\mathbf{i} - \sqrt{2}\mathbf{j})/\sqrt{11}$.

The polarization direction is $(\sqrt{2}\mathbf{i} + 3\mathbf{j} + \sqrt{5}\mathbf{k})/4$.

The amplitude is 4.

The wavelength is $2\pi/\sqrt{11}$.

The frequency is 1.5×10^8 rad/s.

The velocity of propagation $\omega/\mathbf{k} = 1.5 \times 10^8/\sqrt{11}$.

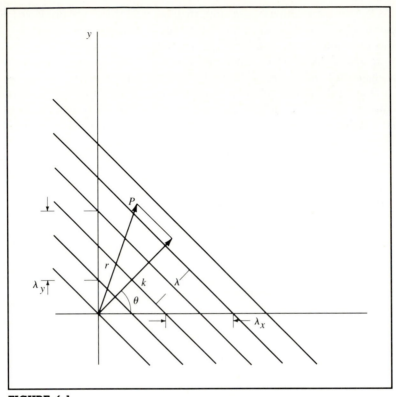

FIGURE 6.1.

The spherical wave arises in a point source radiating uniformly into an isotropic medium. The surfaces of constant phase are concentric spheres centered at the source. Unlike the plane wave, the spherical wave has an amplitude that is position dependent. The form of the spherical wave will be taken as

$$E = E(\mathbf{r})e^{j(\omega t - \mathbf{k} \cdot \mathbf{r} + \phi_0)} \qquad (6.19)$$

The energy of the electromagnetic wave is $\epsilon|\mathbf{E}(\mathbf{r})|^2$, where ϵ is the permittivity of the medium. The energy density of a wavefront at 1 m is given by

$$\mathscr{E} = \frac{\epsilon E_o^2}{4\pi}$$

and at r m by

$$\mathscr{E} = \frac{\epsilon E^2(r)}{4\pi r^2}$$

Conservation of energy applied to the wave requires the energy be equal at both 1 m and r m so that

$$E(r) = \frac{E_o}{r} \qquad (6.20)$$

and the spherical wave has the form

$$E = \frac{E_o}{r} e^{j(\omega t - \mathbf{k} \cdot \mathbf{r} + \phi_0)} \qquad (6.21)$$

A similar argument applied to a cylindrical wave, the wave arising from a uniformly radiation line source, gives

$$E = \frac{E_o}{\sqrt{r}} e^{j(\omega t - \mathbf{k} \cdot \mathbf{r} + \phi_0)} \qquad (6.22)$$

The cylindrical wave decays as $r^{-1/2}$ and thus more slowly than the spherical waves, which decay as r^{-1}, while no position-dependent decay is found with plane waves.

Superposition

The critical property of waves which one needs to understand at this juncture is the *principle of superposition*, the addition of the instantaneous amplitude of two or more waves at a point in space. It is superposition that leads to the phenomena which one observes in interference and diffraction.

Consider two waves whose amplitudes and frequencies are the same but which have directionally different wave vectors and different phases. These two waves are represented by

$$\begin{aligned} \mathbf{E} &= \mathbf{E}_o e^{j(\mathbf{k}_1 \cdot \mathbf{r} - \omega t + \phi_1)} \\ \mathbf{E}_2 &= \mathbf{E}_o e^{j(\mathbf{k}_2 \cdot \mathbf{r} - \omega t + \phi_2)} \end{aligned} \qquad (6.23)$$

and one is concerned with the field in a region where both waves are simultaneously present. The vectors $\mathbf{k}_1$ and $\mathbf{k}_2$ can be decomposed into parallel and normal components as in Figure 6.2:

$$\begin{aligned} \mathbf{k}_1 &= \mathbf{k}_{1p} + \mathbf{k}_{1n} \\ \mathbf{k}_2 &= \mathbf{k}_{2p} + \mathbf{k}_{2n} \end{aligned} \qquad (6.24)$$

Substituting these into (6.23) and superposing the two waves one gets

$$\mathbf{E} = \mathbf{E}_1 + \mathbf{E}_2 = \mathbf{E}_o e^{j(\mathbf{k}_{1p} \cdot \mathbf{r} + \mathbf{k}_{1n} \cdot \mathbf{r} - \omega t + \phi_1)} + \mathbf{E}_o e^{j(\mathbf{k}_{2p} \cdot \mathbf{r} + \mathbf{k}_{2n} \cdot \mathbf{r} - \omega t + \phi_2)} \qquad (6.25)$$

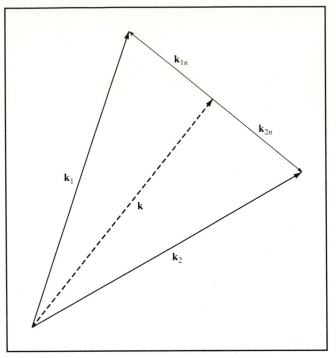

FIGURE 6.2.

Now with

$$\mathbf{k}_{1p} = \mathbf{k}_{2p} = \mathbf{k}$$

and

$$\mathbf{k}_{1n} = -\mathbf{k}_{2n} = \mathbf{k}'$$

substituting these into (6.25) and taking

$$\phi_1 = \phi + \Delta\phi$$
$$\phi_2 = \phi - \Delta\phi$$

one gets

$$\mathbf{E} = \mathbf{E}_o e^{j(\mathbf{k}\cdot\mathbf{r} - \omega t + \phi)}[e^{j(\mathbf{k}'\cdot\mathbf{r} - \Delta\phi)} + e^{j(\mathbf{k}'\cdot\mathbf{r} + \Delta\phi)}]$$

and

$$\mathbf{E} = 2\mathbf{E}_o \cos(\mathbf{k}' \cdot \mathbf{r} + \Delta\phi) \, e^{j(\mathbf{k}\cdot\mathbf{r} - \omega t + \phi)} \qquad (6.26)$$

The result is a position-dependent amplitude which vanishes when

$$\mathbf{k}' \cdot \mathbf{r} + \Delta\phi = \frac{\pi}{2}(2n + 1) \tag{6.27}$$

but which has amplitude $2\mathbf{E}_o$ when

$$\mathbf{k}' \cdot \mathbf{r} + \Delta\phi = n\pi \tag{6.28}$$

The region where the superposition is taking place, called the *interference region,* has a sinusoidally shaped, time-independent amplitude function.

A portion of the amplitude function of the interference region is shown in Figure 6.3. Along the dashed lines the amplitude is always zero and no illumination exists, while along the heavy solid lines the illumination is a maximum.

What will be seen in a region of interference? When a viewing screen is inserted into the interference region, the screen will have alternating bright and dark bars, as shown in Figure 6.4. They will appear constant in time because the energy of the light, which is the quantity that is detected, is given by

$$\mathscr{E} = \frac{\epsilon}{2} E E^* \tag{6.29}$$

where E^* is the complex conjugate of E. Using (6.26) one has for the wave here

$$\mathscr{E} = \frac{\epsilon}{2} E_o^2 \cos^2(\mathbf{k}' \cdot \mathbf{r} + \Delta\phi) \tag{6.30}$$

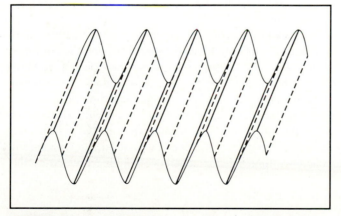

FIGURE 6.3. The amplitude function in a portion of the interference region.

FIGURE 6.4.

Coherence

The assumption made in the preceding section was that a definite phase relationship exists between the interfering waves. Waves which have a fixed relative phase over time are known as *coherent waves*. The reason that two waves derived from a single source were used was to guarantee the constant phase relationship, since a point source is self-coherent. Coherent waves are familiar from the phenomenon of beats in sound, a comparatively easy phenomenon to produce, but they are less common with optical waves.

If each source were a point source, such a source would be coherent. Most sources, however, are *extended;* that is, they have a spatial extent. For such a source to be coherent a definite phase relationship would have to exist between any two points such as A and B in Figure 6.5. Most sources encountered are thermal and produce light by a chemical process like burning or a physical process involving the thermal excitation of a collection of atoms. Both these processes are basically statistical, and the correlation between two points on the source is small. All sources are somewhat coherent and are called *partially coherent*. Purely incoherent light or totally coherent light are extremely rare entities.

For highly coherent sources such as an illuminated pinhole, the coherence properties are described by the coherence time and coherence length. The coherence length is the distance over which the source remains coherent when caused to interfere with itself. This is a relatively direct measurement that will be discussed along with the Michelson interferometer. The coherence time τ is the coherence length ξ divided by the velocity c:

$$\tau = \frac{\xi}{c} \tag{6.31}$$

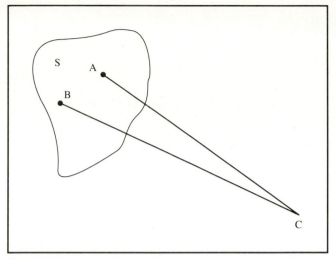

FIGURE 6.5.

The illumination on this page that you are reading, for example, comes from an extended, largely incoherent source. The illumination is quite constant over the surface of the page. Turning on a second light increases the illumination but does not lead to any interference effect. This can be related to the waves generated by the individual sources which are part of the extended source. Consider a point C on the page illuminated by sources at points such as A and B in the monochromatic source S shown in Figure 6.5. The illumination due to the n points making up the source S will be composed of n sources such as

$$E_i = E_o e^{j(\mathbf{k}\cdot\mathbf{r} - \omega t + \phi_n)} \tag{6.32}$$

Letting

$$r_n = r_0 + \Delta r$$

one gets

$$E = \sum E_n = E_o e^{j(\mathbf{k}\cdot\mathbf{r} - \omega t)} \sum_n e^{j(\mathbf{k}\cdot\Delta\mathbf{r} + \phi_n)} \tag{6.33}$$

The source-dependent terms can be grouped with

$$\mathbf{k} \cdot \Delta\mathbf{r} + \phi_n = \zeta_n$$

in which case

$$E = E_o e^{j(\mathbf{k} \cdot \mathbf{r}_0 - \omega t)} \sum_n e^{j\zeta_n}$$

or

$$E = E_o e^{j(\mathbf{k} \cdot \mathbf{r}_0 - \omega t)} \sum_n (\cos \zeta_n + j \sin \zeta_n) \tag{6.34}$$

What one detects on the page is the energy density of the light, or E^*E as in (6.29). Expanding (6.34) one gets

$$\mathscr{E} = \frac{\epsilon}{2} E^* E$$

$$= \frac{\epsilon}{2} E_o^2 \left[\sum_n (\cos^2\zeta_n + \sin^2\zeta_n) + 2\sum_m \sum_n (\cos \zeta_n \cos \zeta_m + \sin \zeta_n \sin \zeta_m) \right]$$

The first term in the brackets is simply n, and the second can be written as the cosine of the difference between angles:

$$\mathscr{E} = \frac{\epsilon}{2} E_o^2 \left[n + 2 \sum_{m>n} \sum_n \cos(\zeta_m - \zeta_n) \right] \tag{6.35}$$

In a completely noncoherent case the double sum contains all possible values of cosine equally and thus sums to zero, and the energy reduces to

$$\mathscr{E} = n \frac{\epsilon}{2} E_o^2 \tag{6.36}$$

In a random source the illumination is a result of the sum of the individual noncoherent point sources and not zero as one might think if one approached the problem by pairing individual sets of coherent terms. Since the period during which each individual source in the parent source is radiating is finite as a result of the decay of individual atomic excited states, it is only self-coherent for the decay period or about 10^{-8} s.

Young's Experiment

The classical example of an interference experiment is that due to Thomas Young, a British physician, in the early 19th century. The experimental arrangement is illustrated in Figure 6.6. Young used a candle as a source, but any bright source will suffice. The single

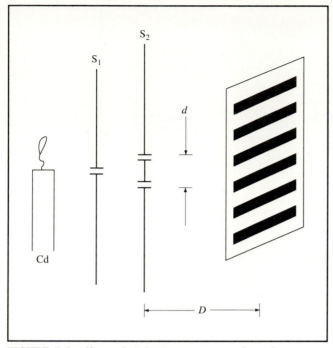

FIGURE 6.6. Young's interference experiment.

slit in screen S_1 (Young used pinholes rather than slits) serves to convert the extended source Cd to a pointlike source. Screen S_2 contains two pinholes or slits separated by a small distance d. For the time being, all slits will be considered to be vanishingly thin. The distance d is typically of the order of a fraction of a millimeter since a wide separation is sufficient to degrade the interference field rapidly as a result of the relatively short coherence length of most incandescent sources.

One can examine first the path of the light from the slits S_2 to the screen. In Figure 6.7a the paths to the center of the screen are shown. The slits S_2 are usually arranged symmetrically with S_1 so that the phases at the two slits S_2 are the same. Clearly both are equal, and the illumination here will be a maximum since there will be no phase difference. At an oblique position, as in Figure 6.7b, the situation is somewhat different. Clearly, the paths from the two slits have different lengths. As a result, the relative phase of the waves reaching the screen along these two paths is different and the illumination at a point P on the screen will depend on that phase difference.

Figure 6.7c offers an enlarged view of the geometry of the problem. The observation point P will be at a vertical displacement y from the midpoint of the slits. The line S_2–Q is drawn as a normal to the line joining the origin O with P. The phase difference between S_1–P and S_2–P is the extra distance S_1–Q divided by the wavelength λ. Note

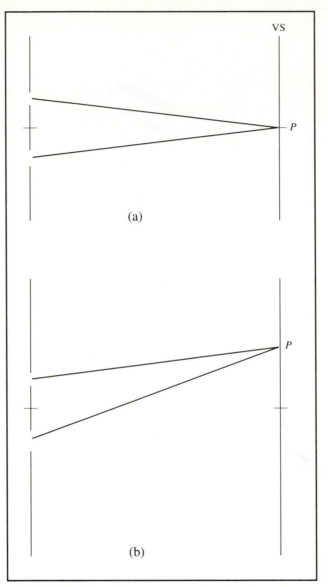

FIGURE 6.7.   The illumination of the viewing screen
VS in Young's experiment. (a) Symmetric and
(b) oblique illumination.

that the angle S_1S_2P is equal to y_0OP since their sides are mutually perpendicular. If
$D \gg y$, this angle is simply y/D, and

$$S_1-Q = d\frac{y}{D}$$

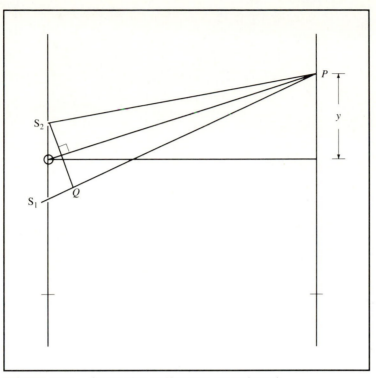

FIGURE 6.7c.   Geometry of Young's experiment.

so that the phase difference will be

$$\Delta_{\text{phase}} = 2\pi \frac{dy}{\lambda D} \tag{6.37}$$

When this is an even multiple of 2π there will be a maximum for

$$m\lambda = \frac{dy}{D} \qquad m = 0, 1, 2, 3, \dots \tag{6.38}$$

while when Δ_{phase} is an odd multiple of π, the illumination at P will be zero:

$$\left(m + \frac{1}{2}\right)\lambda = \frac{dy}{D} \qquad m = 0, 1, 2, 3, \dots \tag{6.39}$$

Equations (6.38) and (6.39) are simply equations (6.27) and (6.28) with the geometrical factors of this particular experimental arrangement. The approximation made for the angle

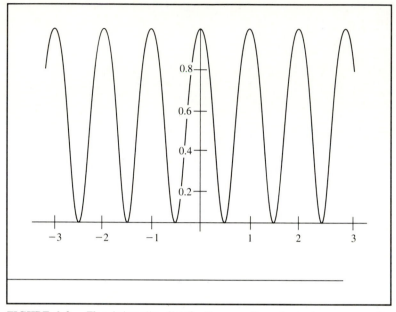

FIGURE 6.8.   The intensity distribution on the observing screen in Young's experiment.

is similar to the paraxial approximation in the geometrical model and has the implication that one should not expect to find the interference pattern far from the central axis. Figure 6.8 illustrates the central intensity distribution on a viewing screen.

EXAMPLE 6.2

Two narrow parallel slits separated by a distance of 0.5 mm are illuminated by a new laser. The viewing screen is at a distance of 0.5 m from the plane of the slit. The fourth maximum is found at a distance of 2.22 mm from the central maximum. What is the wavelength of the laser light?

Solution

Equation (6.37) with $m = 4$ gives

$$4\lambda = \frac{0.5 \text{ mm} \times 2.22 \text{ mm}}{500 \text{ mm}}$$

and

$$\lambda = 0.000555 \text{ mm} \equiv 555 \quad \text{nm}$$

EXAMPLE
6.3

Young's experimental arrangement can be used to monitor the width of the wire leads on transistors. The leads are clamped in a narrow opening as illustrated in Figure 6.9. The slit is then illuminated with highly coherent light. The displacement of the tenth maximum is observed on a screen at 0.5 m. The diameter of the lead is to be held to 0.100 ± 0.005 mm. Between what limits of y must the tenth maximum lie so that the lead will have an acceptable diameter?

Solution

Again one uses equation (6.38). At the nominal diameter of 0.100 mm the 10th maximum will fall at

$$y = \frac{10\lambda D}{d} = \frac{10 \times 627 \times 10^{-6} \text{ mm} \times 500 \text{ mm}}{0.100 \text{ mm}}$$

and

$$y = 31.35 \quad \text{mm}$$

The shift in y with change in d can be found by differentiating, so that

$$\delta y = -\frac{10\lambda D}{d^2}\delta d = -\frac{10\lambda D}{d}\frac{\delta d}{d}$$

and

$$\delta y = 31.35 \text{ mm} \times \frac{0.005 \text{ mm}}{0.100 \text{ mm}} = \pm 1.5675 \quad \text{mm}$$

A high-order maximum, 10 in this case, is used in such situations to give a rather large range for defining an acceptable product. If the first maximum were taken, the range would be 3.135 ± 0.1505 mm, and the pass–fail decision would be more difficult.

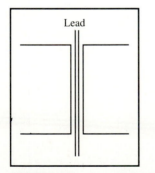

FIGURE 6.9. Example 6.3.

The Michelson Interferometer

The Michelson interferometer was developed by Albert Michelson about 1880. It has since played a premier role in the development of 20th-century physics, particularly in the theory of relativity. It is a highly accurate measuring device and is used to this day.

The schematic diagram of the Michelson interferometer is shown in Figure 6.10. Light from the source S is rendered parallel by lens L_1 and is directed onto a partially silvered mirror PSM. The reflectivity of the PSM is typically 0.5, so that half the light is reflected in arm 1 toward mirror 1 and half is transmitted into arm 2. At the end of each arm the mirror is set to return the light toward the PSM. Half the light returned from M_1 is transmitted through the PSM, and half the light returned from M_2 is reflected at PSM. The light is combined and sent to a detector, which may simply be an eye viewing through a telescope represented by L_2.

A plate made of the same material as the PSM but without the partially reflecting surface is set in arm 1 as a *compensating plate* to ensure that the light path in each arm is rendered equivalent by having the light in each arm pass through the same thicknesses of the same materials. Mirror M_2 is equipped with a micrometer screw Sc so that M_2 may be

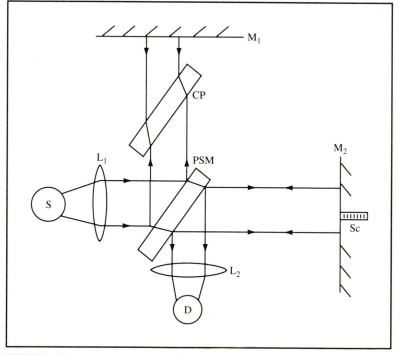

FIGURE 6.10.   The Michelson interferometer. The detector D is often the eye.

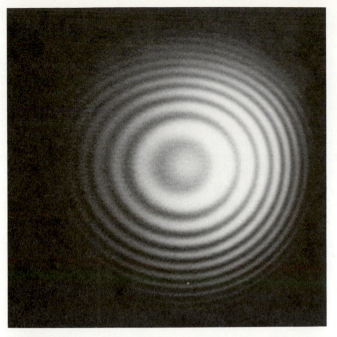

FIGURE 6.11a. The circular fringes observed in a Michelson interferometer.

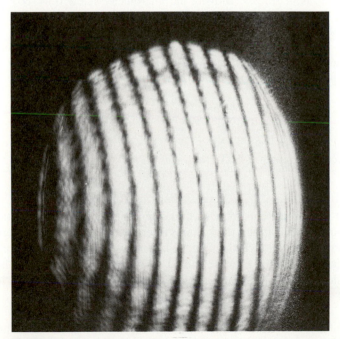

FIGURE 6.11b. The fringes observed in a Michelson interferometer with a tilted mirror.

moved forward and backward along the arm, thereby adjusting the length of arm 2 relative to arm 1. The micrometer scale allows one to determine accurately the distance of traverse of the mirror.

When the observer at D in Figure 6.10 views the light from the PSM, he sees a set of concentric ring fringes as in Figure 6.11a. When the paths are strictly equal, the center spot will be bright because of constructive interference in the on-axis paths. As one scans toward the periphery of the field, successive dark and bright fringes will be seen, as in the figure.

The Michelson interferometer is not usually used in the configuration shown, with mirrors normal to the light paths in the arms. The mirrors are usually tilted slightly by means of adjusting screws provided for that purpose. The effect of the tilt is to move the center of the circular fringe from the center of the viewing field to a position well removed from that center. The interference fringes then appear as nearly straight lines in the viewing field as in Figure 6.11b. Moving the adjusting screw Sc in Figure 6.10 will cause the fringes to sweep through the field. Each movement of the micrometer screw by 1/2 wavelength will effect a 1-wavelength change in the total path in the arm and will move the bright fringe at the center of the field one place and replace it with the next adjacent bright fringe. In this way, for example, the wavelength of a source may be measured.

EXAMPLE 6.4

A bright monochromatic source is used as a source for a Michelson interferometer adjusted to give straight fringes. Moving the micrometer screw 0.04166 mm causes 150 bright fringes of the interference pattern to sweep past the center line of the viewing field. What is the wavelength of the source?

Solution

The 150 bright fringes represent a 150/2 or 75-wavelength displacement of the mirror. The mirror displacement is 41.66 μm and

$$\frac{41.66}{75} \ \mu m = 0.5555 \ \ \mu m$$

and the source wavelength is 555.5 nm.

EXAMPLE 6.5

A Michelson interferometer is to be used to determine the index of refraction of a plastic film. To do this, the 0.040-mm-thick film is inserted in arm 1 of the interferometer so that half the beam passes through the film. When viewed from the detector position, the fringes passing through the film are shifted as shown in Figure E6.5. The interference pattern has shifted by 83.6 bright fringes in the region of the film. What is the index of refraction of the film? The light in the interferometer has a 555-nm wavelength.

Solution

A shift of 83.6 bright fringes implies that the total path through the film in arm 1 exceeds the unencumbered path by 83.6 wavelengths. Note that the total thickness of plastic film through which the light travels is 0.080 mm or 80 μm since the film is traversed once in

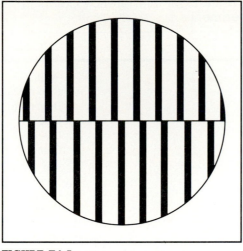

FIGURE E6.5.

each direction along the arm. The wavelength of the light within the film is the wavelength in air divided by the index of refraction n of the film, the quantity to be found. The number of wavelengths in the film is the number in air plus 83.6. The total number of waves in the film is

$$\frac{80\ \mu\text{m}}{0.555\ \mu\text{m}} + 83.6 = 227.744$$

and the wavelength in the film is

$$\frac{80}{227.744} = 0.35127\ \mu\text{m} \equiv 351.25 \quad \text{nm}$$

This is the wavelength in air divided by the index of the film

$$\frac{\lambda_{\text{air}}}{n} = 351.17 \quad \text{nm}$$

and $n = 1.5799$.

This example introduces an interesting quandary. When all the fringes at 555 nm appear the same, how can one fix the shift at 83.6 fringes? In fact, in the problem as posed, one cannot. As the film is inserted into the beam, the shift is immediate, not gradual, and there is no way to count the shift as the film is inserted. How can one approach this problem? The answer lies in using white light rather than a monochromatic source.

TABLE 6.1 White-Light Interference Colors in Young's Experiment

$\Delta_{\text{Path}}(\mu m)$	Color
0	White
0.158	Brown-white
0.259	Bright red
0.332	Blue
0.565	Green
0.664	Orange
0.747	Red
0.866	Violet
1.101	Green
1.376	Violet

When white light is used in the Michelson interferometer with the arms set strictly equal, all wavelengths will interfere constructively and the observer will see the white source. As the length of one arm varies with respect to the other, some wavelengths will interfere constructively while others will interfere destructively. Only when the paths are equal will the fringe be white. Actually, the fringes first seen by Young were not the simple bright and dark bands but an array of colors. Table 6.1 lists the colors observed as a function of the path difference.

As one causes one arm of the interferometer to lengthen relative to the other arm, the intensity of the colors fades and the field eventually reverts to a white-light field with no visible fringes. The reason the fringes vanish is that the rapid juxtaposition of different-colored maxima when high orders of interference are present is sufficient to give the impression of a white-light field to the eye. One can typically see about six orders in white light compared with several hundred in monochromatic light.

The white-light fringe system can be used to alleviate the problem encountered in Example 6.5. Insertion of the film in the light path will shift the central white-light fringe from the center of the field for the light passing through the film. The micrometer screw in arm 2 can be used to make the light paths equal to the path in arm 1, which includes the film. In so doing the white-light maximum in the portion of the arm not including the film will be shifted away from the center of the field. The change in optical path length can then be read from the micrometer.

EXAMPLE 6.6

What change in optical path length will be read from the micrometer for the film given in Example 6.4 when the interferometer is initially set with equal arms and is then readjusted to a central bright fringe after the film has been inserted?

Solution

The change in optical path is the additional number of wavelengths in the arm with the film. The optical path length in air is 80 μm while in the film it is 80 μm × 1.5799 or 126.39 μm. The adjustment to the micrometer will be the difference between these or 46.39 μm.

Visibility of Fringes

If one takes a highly monochromatic source and extends one arm relative to the other, the fringes will decrease in intensity and eventually vanish. This is a result of the coherence properties of the source. In a thermal source the light is generated by the decay process in the excited states of individual atoms. Typically, the decay of an atom takes place in about 10^{-10} s; the length of the wave from an individual decay will be given by

$$\delta_c = 3 \times 10^8 \frac{m}{s} \times 10^{-10}\ s = 0.03 \quad m$$

or about 3 cm. When the length of one arm of the interferometer then approaches ~ 1 cm difference in length compared with the other arm, the fringes fade in intensity; their *visibility* decreases.

The visibility of fringes is defined by

$$V = \frac{I_{max} - I_{min}}{I_{max} + I_{min}} \tag{6.40}$$

When the fringes are at full intensity in the bright bar and zero intensity in the dark bar the visibility equals one, as would be the case in Figure 6.12a. As the coherence of the source decreases, one has a reduced visibility as in (b) and (c), until finally, as the coherence vanishes, I_{max} and I_{min} converge and V goes to zero. Various definitions of partial coherence are based on the visibility of fringes.

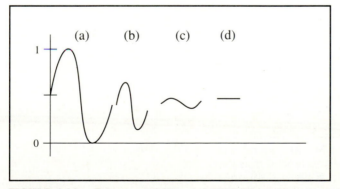

FIGURE 6.12.   Fringe visibility. (a) Visibility unity. (b, c) Visibility decreasing. (d) Zero visibility.

Polarization

Equation (6.17), the solution to the spatial part of the wave equation, has a vector for its amplitude. In the sequel to (6.17) this was taken to lie in one direction to facilitate the discussion of interference. The electric field amplitude is a transverse vector, however, with the magnetic field amplitude a second vector normal to the electric field, and both of these are normal to the direction of propagation as expressed by the wave vector **k**. This is shown in Figure 6.13. When the electric field amplitude vector is constrained to lie in one direction only, the wave is said to be *linearly polarized*. Generally, light from a thermal source is randomly polarized, but a single direction of polarization can be selected and imposed upon the propagating wave using *polarizing filters*, which are readily available commercially. Once the light has passed through such a filter it is plane polarized.

An interesting experiment involving polarization can be performed using Young's apparatus as discussed earlier. If polarizing filters are placed over the slits with their polarizing axes parallel, then there is no change in the pattern observed on the screen. If, however, one of the filters has its axis perpendicular to that of the other, the interference pattern on the screen vanishes. This, in fact, was essentially the experiment done by Young from which he concluded that light was a transverse wave. In general, with randomly polarized light one has interference since all the possible polarization orientations are present in the light so that those that are parallel interfere with others with the same orientation.

The commercial filters mentioned earlier are commonly called *Polaroid filters*. They are produced by preparing plastic sheeting with particular long-chain molecules in

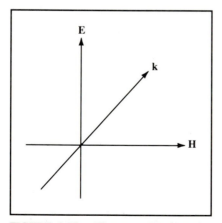

FIGURE 6.13.   The amplitude vectors of the electric field **E** and the magnetic field **H** lie at right angles to the propagation vector **k**.

them and then stretching the sheet to align the molecules. One can then imagine a picket fence effect where only waves parallel to the slats can traverse the "fence." This is illustrated in Figure 6.14a, where the randomly polarized waves leave a residual linearly polarized wave after traversing the filter; the random assortment of **E** vectors prior to crossing the filter is converted into a single **E** amplitude vector parallel to the filter axis.

If one now passes this linearly polarized light through a second polarizer, the intensity of the light after the second filter, called the *analyzer,* is related to the incident intensity by *Malus's law*

$$I = I_\mathrm{m} \cos^2 \theta \qquad (6.41)$$

where I_m represents the maximum transmitted intensity and θ is the angle between the axes of the polarizing and analyzing filters. Malus's law was discovered at about the same time as Young's classical experiment and clearly establishes the transverse nature of light waves since such an effect could not occur with longitudinal waves.

Since $I = \mathbf{E} \cdot \mathbf{E}$, the implication of Malus's law is that a wave falling on the analyzer with its amplitude vector at an angle θ to the polarizing direction can be decomposed into components normal to and parallel to the polarizing axis and that only those components parallel to the polarizing axis are passed by the filter. The geometry is illustrated in Figure 6.14b.

Polarization effects are important in fiber optic systems and in laser systems. In addition, analysis of the polarization properties of scattered light gives important informa-

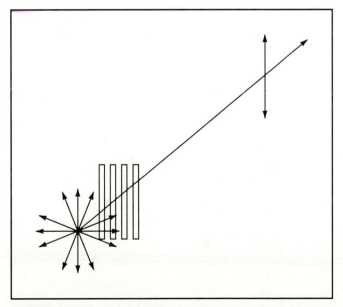

FIGURE 6.14a. Production of a linearly polarized wave with the "picket fence" of a Polaroid filter.

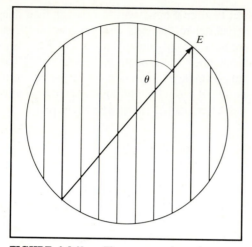

FIGURE 6.14b. The geometry of
Malus's law.

tion about the nature of the scattering body. This scattering polarization has been applied
to studies ranging from Saturn's rings and cosmic dust to the size and shape of viruses.

Polarization of a light wave can be produced in a number of ways, including scattering when the light is reflected from a surface. In reflection and the concomitant refraction
at a surface, the components of the light incident on the surface with polarization parallel
to the surface behave differently from those whose polarization directions intersect the

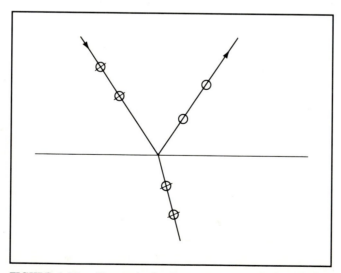

FIGURE 6.15. The polarization components parallel to
the surface are represented by the open circles, while
those components in the plane of the figure are represented by crossed circles.

surface. Figure 6.15 illustrates this. The light wave incident on the surface is randomly polarized and can be represented by its components, one parallel to the surface and the normal components lying in the plane of the figure. After reflection and refraction at the surface, the reflected light contains primarily the component parallel to the surface. Glare-reducing Polaroid sunglasses take advantage of this effect by having the filters in the lenses set with their polarizing axes vertically. There is an angle, *Brewster's angle*, at which only the parallel component is reflected and this angle is given by *Brewster's law*

$$\tan \theta_p = n \qquad\qquad (6.42)$$

where n is the index of refraction of the reflecting object.

EXAMPLE 6.7   What is the Brewster angle for light incident on plate glass, $n = 1.51$?

Solution   Using Brewster's law (6.42),

$$\tan \theta_p = 1.51$$

$$\theta_p = 56.48°$$

corresponding to an angle of refraction of 33.52°. Note that the sum of the angle of reflection and the angle of refraction is 90°, so that the reflected and refracted rays are at 90° to each other when the incidence is at Brewster's angle (Figure E6.7).

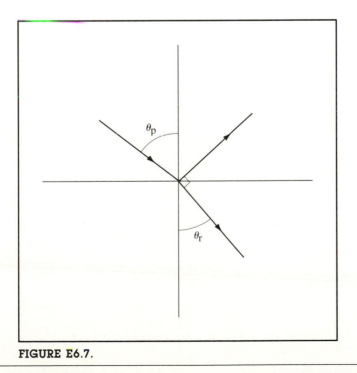

FIGURE E6.7.

Brewster's law can be used to produce linear polarization from randomly polarized light. A stack of glass plates are set at the Brewster angle θ_q with reference to the direction of polarization. Each successive interface removes a fraction of the remaining light polarized parallel to the surface. After passing seven or eight plates, the transmitted light is nearly completely polarized.

Certain crystals, notably naturally occurring calcite and quartz, are optically anisotropic; that is, they have different optical properties along different crystalline axes. A beam of unpolarized light falling normally on one face of a calcite crystal will emerge from the opposite face as two beams, each of which will be linearly polarized at right angles to the other. The rays are labeled ''o'' and ''e'' for ordinary and extraordinary. The o-ray obeys Snell's law but the e-ray does not; in fact, the e-ray may not even lie in the plane of incidence. Among the crystals that exhibit this effect are ice and quartz, but with these crystals, the effect is not nearly so pronounced as with calcite. The difference in refractive index between the o-ray and the e-ray for ice is 0.004 and for quartz 0.012, while for calcite it is -0.172. Materials which are doubly refracting are called *birefringent*.

An interesting device can be produced from a slab of calcite cut so that the emerging e-ray and o-ray have a path difference of $\lambda/4$. Such a plate is called a *quarter-wave plate*, and the emerging light is said to be *circularly polarized*. The two plane waves emerging from the second face can be represented by phasors rotating at an angular velocity ω and at $90°$ to one another, where is the angular frequency of the light wave. Thus at different times the wave appears to be polarized along different directions.

EXAMPLE 6.8   What is the thickness of a quarter-wave plate of calcite for use at 590 nm? At 590 nm, $n_e = 1.486$ and $n_o = 1.658$.

Solution   If the crystal thickness is taken to be t, the number of wavelengths of the extraordinary ray in the crystal is given by

$$N_e = \frac{tn_e}{\lambda}$$

and for the ordinary ray

$$N_o = \frac{tn_o}{\lambda}$$

The difference between these should be one-fourth:

$$N_o - N_e = \frac{1}{4} = \frac{t}{\lambda}(n_o - n_e)$$

This gives

$$t = \frac{\lambda}{4} \frac{1}{\Delta n} = \frac{590}{4 \times 0.172} = 858 \quad \text{nm}$$

The thickness is 0.858 μm, and it is not feasible to cleave a calcite plate to this thickness. A quarter-wave plate is more easily produced from mica, for which Δn is 0.012, corresponding to a quarter-wave plate thickness of 12 μm, which can be gotten more easily by peeling a very thin layer from a mica sample.

The difference between the indices of refraction of the o-ray and the e-ray may be exploited to produce linearly polarized light. If a prism is cut so that a ray entering normal to a face encounters the reflecting face at an angle greater than the critical angle, it will be totally reflected. For calcite, the critical angles are $37.09°$ for the o-ray and $42.29°$ for the e-ray. For quartz at 700 nm, near the output wavelength of the ruby laser, the critical angles are $40.2°$ and $40.5°$. The prism is made in two parts with an air gap as shown in Figure 6.16. The second prism is used to keep the transmitted ray parallel to the incident ray. The prisms must be cut with the crystal axes as shown in Figure 6.16. In the first prism the axis is normal to the page, and in the second prism it is parallel to the page. This device is known as a *Glan–Thompson prism* and finds use in low- to medium-power laser systems.

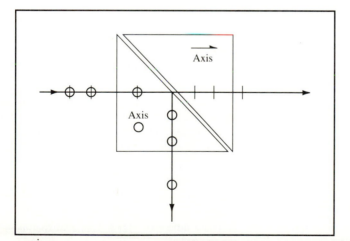

FIGURE 6.16. The Glan–Thomson prism.

PROBLEMS

6.1. Using a method identical with that for the **E** field, show that **H** satisfies the wave equation with the same velocity of propagation as **E**.

6.2. The current density in a medium is given by $\sigma\mathbf{E}$, where σ is the conductivity. Given a wave $\mathbf{E} = \mathbf{E}_o e^{j\omega t}$ propagating in a medium with conductivity σ and $\rho = 0$, show that the wave is attenuated.

6.3. A plane wave is given by

$$\mathbf{E} = 50 \sin(10^8 t + 2ax)\hat{\mathbf{j}}$$

Find the direction of propagation, λ, ω, f, and **H**.

6.4. Which of the following can represent waves:
   (i) $30e^{j\omega(t-3x)}$
   (ii) $\sin \omega(5t + 10z)$
   (iii) $x^2 + 4xt + t^2$
   (iv) $\sin^2(2\pi t + \Delta\phi)$
   (v) $\sin t \cos z$
   (vi) $\sin(5x + 6y)$
   (vii) $r^{-3}e^{j\omega t}$

6.5. A plane wave propagating in the z direction has amplitude 8, frequency 10 Hz, wavelength 4 m, and magnitude 4 at the origin when $t = 0$. Write the equation for the wave.

6.6. Show that equations (6.10) and (6.11) are equivalent.

6.7. Why does $(\hat{\mathbf{j}} + \hat{\mathbf{k}})e^{j(\omega t + 3y + 3z)}$ not represent an optical wave?

6.8. Show that the amplitude of a cylindrical wave decays as $r^{-1/2}$.

6.9. In Young's experiment the slit separation is 100 μm, and the illuminating light is at 555 nm. At what angle is the first maximum? the tenth maximum? the fifth minimum?

6.10. Design an interferometer such that the maxima will be 1.5° apart on the observing screen for 555-nm light.

6.11. An interferometer of Young's type with its slits 120 μm apart is used with green 546-nm light and the interference pattern is observed on a screen 1.0 m from the slits. A thin plastic sheet in which the optical path is 0.5 wavelength longer than the initially equal air path is placed over one of the slits. Describe the fringe shift, and specify the distance the fringes move on the screen.

6.12. Two sources, one at 555 nm and a second at 660 nm, illuminate the two slits of a Young's interference experiment. Will any of the maxima of the two interference patterns coincide, and if so, under what circumstances?

6.13. Describe the trajectory of equal phase differences for a Young's experiment.

6.14. A mirror is unaffected by a change in the external medium. If one were to change the medium in which a Young's interference pattern was being observed, what changes, if any, would occur?

6.15. The moving mirror in a Michelson interferometer is moved through 247 μm and an 853-fringe shift occurs. What is the wavelength of the light?

6.16. Polarizing filters are placed in each arm of a Michelson interferometer. What will one observe as one arm is lengthened if their polarization axes are parallel? If their axes are perpendicular? If one is at 45° to the other?

6.17. A thin plastic film is placed in one arm of a Michelson interferometer being used with a 546-nm source. If the index of the film is 1.43 and a 7.25-fringe shift is observed, what is the thickness of the film?

6.18. An unpolarized light wave passes through two polarizing filters, and the intensity observed is 0.4 of that observed without the filters. At what angle are the filters?

6.19. A stack of six polarizing filters are set so that the polarizing angle of each is altered by 0.1 rad with respect to the previous one. What fraction of the incident unpolarized light emerges from the stack? Must each filter be altered in the same sense, say clockwise?

6.20. Describe how one might use a series of polarizing filters to rotate the plane of plane-polarized light. Is this a feasible process?

6.21. At what angle is the light reflected from the surface of a puddle completely polarized?

6.22. A thin section of mica is to be used to form a half-wave plate from 546-nm light. How thick should it be?

6.23. Partially polarized light is passed through a polarizing filter. Describe the intensity of the transmitted wave.

CHAPTER 7

Applications of Interferometry

Interferometry provides a measurement technique that is applied in many technical areas. Measurement precision with interferometric techniques is of the order of a fraction of the wavelength of light, providing a sharpness of resolution not easily duplicated by other approaches. In this chapter several types of measurement effected by interferometric techniques will be discussed. They should be viewed as a sampling in a much larger universe of measurements from which they have been selected, and are intended to give an overview of interferometric measurements rather than a complete catalog. As one works through this chapter, one should attempt to construct other scenarios where similar interferometric measurements would be applicable.

The two interferometers presented in Chapter 6 represent the two common classes of interferometers. These classes are distinguished by the way in which the coherent wavefront is used to create the interference pattern. Young's experiment selects two parts of the wavefront and brings these two selected parts together; this class is called *wavefront division,* since different portions of the same wavefront produce the interference pattern. In contrast, the Michelson interferometer uses a half-silvered mirror and, rather than selecting different portions of the wavefront, uses the entire wavefront; the interference arises because of *amplitude division,* which is the second class of interferometer. This chapter will examine results from other instruments in each of these classes.

Wavefront Division Interferometers

The Rayleigh Interferometer

The Rayleigh interferometer is a wavefront division interferometer used to measure the index of refraction of gases and liquids. The interferometer has two chambers set along the paths of the light exiting the two slits used in this experiment. Each chamber is equipped with a valve that can be used to empty and refill its contents. The source is at S_o, and it illuminates the source slits S_1 and S_2. The light then passes through the two chambers and through two compensating plates and is focused at the detector D. One compensating plate labeled Pc in Figure 7.1 can be rotated while Pf is maintained in a fixed position. The light exiting the system is gathered by a lens and focused at the detector D. The detector may simply be an eye, or it may be a photoelectric system.

In operation, both chambers are evacuated and the compensating plate Pc is rotated so that the detector has the central maximum falling on it. This then calibrates the system and takes into account any differences in the two arms of the interferometer. White light is used in this system. A gas whose index is to be determined is then introduced into one arm of the interferometer, and Pc is again rotated to give a central white fringe. The rotation of Pc gives the change in path length in that arm. Typically, the indices of refraction of gases are small so that the rotational angles of Pc are small.

The principle of operation of the compensating plate is illustrated in Figure 7.2. As the plate rotates, the path length of a ray through the plate increases. Rotating the plate through an angle ϕ gives

$$\sin \theta' = \frac{\sin \phi}{n_p} \tag{7.1}$$

where n_p is the refractive index of the plate. The increase in path length in the plate, Δl, is then

$$\Delta l = t\left[\left(1 - \frac{\sin^2 \theta}{n_p}\right)^{-1/2} - 1\right] \tag{7.2}$$

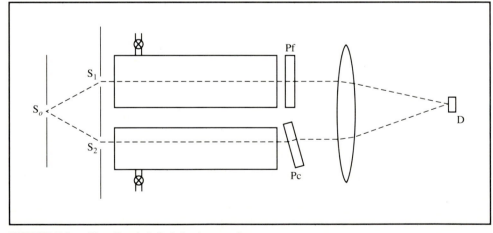

FIGURE 7.1.   The Rayleigh interferometer.

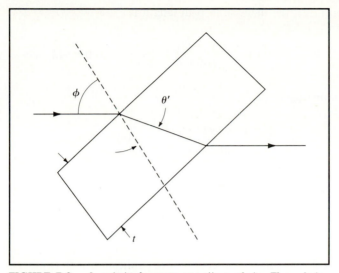

FIGURE 7.2. A rotated compensating plate. The plate thickness is t, and the angle of rotation of the plate is ϕ.

and the phase change $\Delta\phi$ introduced by the plate is

$$\Delta\phi = \frac{\Delta l}{\lambda_o/n_p} = \frac{\Delta l}{\lambda_o}n_p \tag{7.3}$$

The phase change in arm 1 due to the introduction of the gas into the tube is given by

$$\Delta\phi = \frac{L}{\lambda_o/n} - \frac{L}{\lambda_o} = \frac{L}{\lambda_o}(n-1) \tag{7.4}$$

where L is the known length of arm 1. Equating the phase differences given in (7.3) and (7.4) gives

$$(n-1) = \frac{\Delta l}{L}n_p \tag{7.5}$$

EXAMPLE 7.1

The Rayleigh interferometer is used to measure the refractive index of an inert gas. The compensating plate has an index of 1.6000 and is precisely 1.000 cm thick. After the gas has been introduced into arm 1, the compensating plate must be rotated through 10° to return the central white fringe to its original position. Arm 1 has a calibrated length of 29.2 cm. What is the refractive index of the gas?

Solution Using equation (7.2), the increase in path length is

$$\Delta l = 1.00 \left[\left(1 - \frac{\sin^2 10°}{1.6000^2} \right)^{-1/2} - 1 \right] = 0.005842 \quad \text{cm}$$

The index is found using (7.5):

$$n - 1 = \frac{0.005942 \text{ cm}}{29.2 \text{ cm}} \times 1.6000 = 0.000326$$

and

$$n = 1.000326$$

Wavefront division has been seen to have application with two beams. If one considers the use of a Young's type experiment to measure the wavelength of a monochromatic source, one can see that there are several potential sources of error. The separation of the slits as well as their finite width makes the precise determination of d difficult. The centers of the maxima are similarly difficult to determine precisely. An alternative does exist, however, and that is the use of multiple slits.

Consider a system with three slits, where again the separation of the slits is d. The system that will be used is illustrated in Figure 7.3. The experiment will be described in terms of the angle θ the light makes with the normal to the screen containing the slits. Figure 7.4 shows that the phase shift for the third slit is twice that for the second slit relative to slit 1. The amplitude at the screen at an angle is given by

$$A_0(e^{j0} + e^{j\theta} + e^{j2\theta})$$

and the intensity by

$$I_0(e^{j0} + e^{j\theta} + e^{j2\theta})^2$$

The effect of an increasing θ can be seen using phasor diagrams. When $\theta = 0$, all three waves have the same phase and one has the central maximum. As θ increases, the phasor associated with slit 3 rotates at twice the angle associated with slit 2. When $\theta = 120°$, the phasor diagram is as shown in Figure 7.5, where slits 2 and 3 are referenced to slit 1. The result is a zero intensity at 120°. The intensity then increases, again giving a maximum for $\theta = 180°$, as in Figure 7.5b, but this maximum will have an intensity only one-ninth that of the central maximum. Further increase in θ gives a second zero at $\theta = 240°$ as in Figure 7.5c. Finally, at $\theta = 360°$ all three phasors are again aligned and one has a principal maximum equal to the central maximum. The interference pattern on a screen inserted in the field will have an intensity distribution similar to that shown in Figure 7.6.

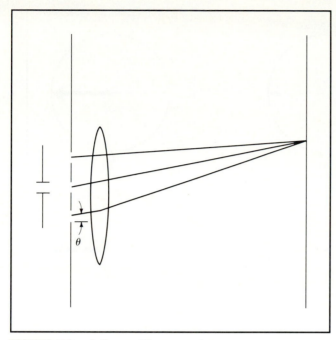

FIGURE 7.3. A three-slit apparatus.

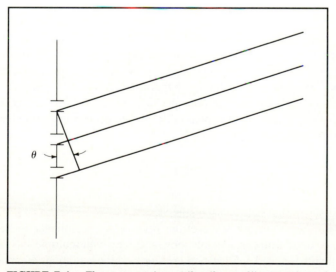

FIGURE 7.4. The geometry of the three-slit experiment.

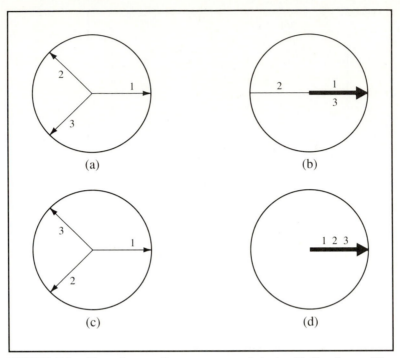

FIGURE 7.5. The phasor diagrams associated with the three-slit experiment. (a) $\theta = 120°$. (b) $\theta = 180°$. (c) $\theta = 270°$. (d) $\theta = 360°$.

One can increase the number of slits beyond three. With four slits one will find three minima and two intermediate maxima, each having 1/16th the intensity of the principal maxima. A further increase in the number of slits will result in an increase in the number of minima between the principal maxima, there being $(n - 1)$, where n is the number of slits. More significant still is the reduction in intensity of the intermediate maxima, which have intensity $1/n^2$ times the intensity of the principal maxima, where again n is the number of slits. When n is very large, of the order of 10^3–10^4 slits, the intermediate maxima are no longer seen, and the principal maxima are widely spread. The condition for the principal maxima is that the phase difference between adjacent slits is $2\pi n$. The principal maxima occur at

$$\sin \theta = n\frac{\lambda}{d} \tag{7.6}$$

where d is the separation between the slits.

If one now wants to use a wavefront division interferometer to measure wavelengths, it can be done with considerably more accuracy with multiple slits. A plate with a high density of slits is called a *diffraction grating,* and these gratings are readily available at a reasonable cost. Their primary function is to provide measurements of the wavelength of radiation.

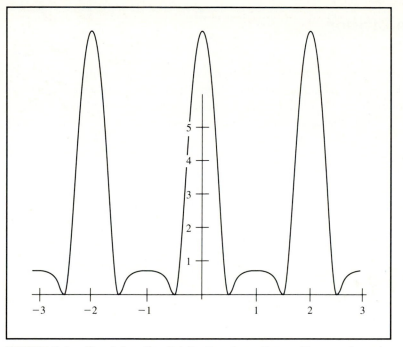

FIGURE 7.6. The intensity distribution for a three-slit experiment.

EXAMPLE 7.2   A grating with 5×10^3 slits per centimeter is used to measure the wavelength of the output of a laser. The output radiation is directed onto a narrow slit and the radiation from the slit is used to illuminate a grating. Maxima are observed using an accurate goniometer to measure the angles of the maxima relative to the central maximum. Maxima are found at angles of $\pm 18.45°$ and $\pm 39.25°$. What is the wavelength of the laser output?

Solution   The slit separation d is given by (1 cm $\approx 10^7$ nm):

$$\frac{10^7 \text{ nm}}{5 \times 10^3 \text{ slits}} = 2 \times 10^3 \quad \text{nm}$$

$$\lambda = \frac{d \sin \theta}{n} = \frac{632.95}{1} = \frac{1265.41}{2}$$

Averaging the two measurements gives $\lambda = 632.83$ nm.

Wavefront division will be treated again in Chapter 8 when diffraction is introduced.

Amplitude Division

The Michelson interferometer exemplifies the amplitude division class of interferometer. The input light has its amplitude divided and subsequently recombined, which allows one to evaluate the differences in the paths in the arms of the interferometer.

An interesting application of the Michelson interferometer involves the calibration of a stepping-motor-driven stage. This stage undergoes a linear displacement as a function of the electrical driving signal applied to the stepping motor. The calibration of the displacement can be evaluated using white-light fringes as described in Chapter 6. A mirror is mounted on the movable stage and this mirror is set in one arm of the interferometer, which is then set to a balanced, equal-arm position. As the signals are applied to the stepping motor, the mirror is displaced and the optical length of the arm of the interferometer with the stepping-motor-driven mirror is changed. Adjustment of the micrometer screw in the other arm to bring the white central fringe to the center of the field gives the value of the displacement of the stage. In this fashion the stepping motor can be calibrated.

An alternative application of the Michelson interferometer is as a control device. If the detector is a photodetector that puts its output into a pulse counting circuit, the motion of the arm can be accurately controlled by the number of fringes counted in the circuit. Very accurate lathes and milling machines are controlled in this way, using one arm along the axis, whose length is to be controlled, and a second reference arm. The position of the movable arm is fixed in terms of a computed number of fringes. Extremely accurate machining can be done in this way.

There are many alternatives to the Michelson interferometer. One of the most common is the Mach–Zender interferometer, as illustrated in Figure 7.7. Here two half-

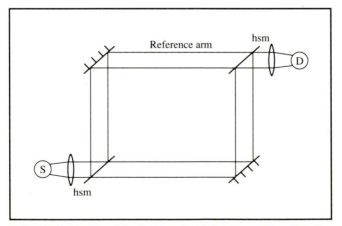

FIGURE 7.7.   The Mach–Zender interferometer. S is the monochromatic source, D the detector, and hsm half-silvered mirrors to divide and recombine the amplitude of the light wave.

silvered mirrors are used, the first as an amplitude divider and the second to recombine the light. Objects to be examined are placed in the experimental arm, and the phase delay is observed at the detector. The uses range from direct observation of the heating of the air flowing over the wing of an aircraft, where the experimental arm traverses a wind tunnel, to observations of inhomogeneity in a glass sample. One use of current importance is in the examination of the crystal rods used as the active medium in lasers. Each rod is inserted in the experimental arm of the interferometer, and index inhomogeneities, if they exist, show as spots in the interference field. The Mach–Zender configuration also finds application as a light modulator in communication systems using light.

The number of forms of interferometer derived from the Michelson is essentially limitless. For example, Figure 7.8 illustrates the Sagnac or cyclic interferometer first used in experiments similar to the Michelson–Morley experiments. Today it is used in ring lasers and as a rotational sensor in much the same way as gyroscopes are used for guidance. As the light travels around the interferometer, which is assumed to be rotating about its center, classically, the light traveling in the direction of rotation will have a longer time of flight around the path than the light traveling in opposition to the direction of rotation.[1] The result will be a shift in the fringe pattern as seen by the observer. This is given by

$$\Delta N = \frac{4A\omega}{c\lambda} \tag{7.7}$$

where ΔN is the number of fringes shifted, A is the area of the interferometer, ω is the angular velocity about the center of rotation C in Figure 7.8, c is the velocity of light, and

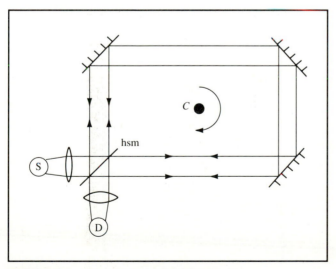

FIGURE 7.8. The Sagnac interferometer.

---

[1] Strictly, one must analyze this system relativistically, but the result turns out to be the same as the classical one.

λ is the wavelength of the monochromatic (laser) source. These interferometers are now being used in inertial guidance systems.

Reflection Revisited

There are a number of reflective surfaces involved in each of the amplitude division systems. If one looks at Figures 6.10, 7.7, and 7.8, one can see that the number of reflections in each arm of the interferometers is the same, so that there is a matching compensation of the reflections. The laws governing reflection and transmission at a boundary surface between two different media are based on the properties of the electromagnetic field at boundary surfaces and depend on the polarization state of the incident wave as well as on the angle of incidence.

For light incident normally at an interface between two optically transparent media, two important results arise. The *reflection coefficient,* the ratio of the incident to reflected amplitude, is given by (see Appendix IV)

$$r = -\frac{n_1 - n_2}{n_1 + n_2} \tag{7.8}$$

and the reflected wave has its phase advanced by π if $n_2 > n_1$.

Two experiments serve to illustrate this. The Lloyd's mirror experiment is a modified Young's two-slit experiment. The experimental arrangement is as shown in Figure 7.9. The source S′ is a virtual source arising from the reflection at the mirror M. The mirror itself is generally a black glass since a metallized mirror surface changes the

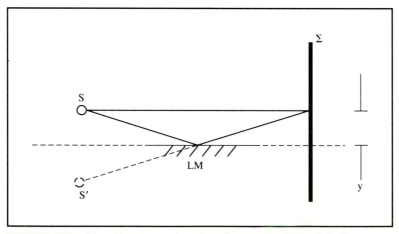

FIGURE 7.9. Lloyd's mirror showing the secondary virtual source S′.

boundary conditions for reflection. What one finds is that the equations for bright fringes and dark fringes (6.37) and (6.38) are reversed and the values of y for which bright fringes are found is

$$y = \frac{D\lambda}{d}\left(m + \frac{1}{2}\right) \tag{7.9}$$

which is easily explained by the phase advance in the path reflected from the mirror.

A second example of this is found with Newton's rings, a procedure of importance in the evaluation of the surface curvatures of lenses as well as in illustrating the phase shift. The layout as used in many optical shops is illustrated in Figure 7.10a. The monochromatic source is usually an uncoated fluorescent tube, and tracing paper can be used as a diffuser. The entire apparatus is set in a box whose walls are blackened to avoid stray reflections. A plate of glass set at 45° provides sufficient intensity in the reflected light for the observer. The typical pattern is shown in Figure 7.10b. The central region is dark. This region arises from the light reflected on the bottom surface of the lens and from the top surface of the test flat. The light reflected within the lens at the interface, a vanishingly thin air film, has $n_1 > n_2$ and does not undergo a phase advance, while the light reflected on the test flat is reflected at an air–glass interface with $n_2 > n_1$ and has a phase change of

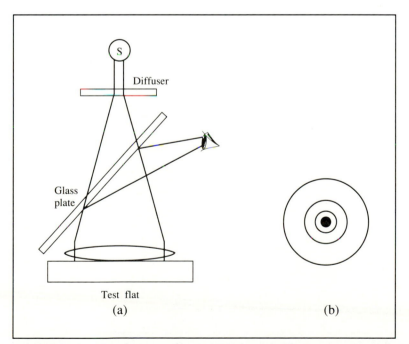

FIGURE 7.10. Newton's rings. (a) The experimental arrangement for a lens test. (b) The appearance of Newton's rings in monochromatic light.

π. The returned light then consists of two reflected waves, one of which is one-half wavelength out of phase with the other, thus interfering destructively.

As one moves away from the center, there is a path difference as well as a phase difference. The path difference is determined by the air gap. The gap is found using the *sagittal theorem,* as illustrated in Figure 7.11. The gap g at a radial distance l from the central contact point can be found using the Pythagorean theorem

$$R^2 = (R - g)^2 + l^2 \tag{7.10}$$

from which

$$g = R - \sqrt{R^2 - l^2} \tag{7.11}$$

The phase shift due to the gap is $2g/\lambda$, and one has dark fringes whenever

$$\frac{2g}{\lambda} = m \qquad m = 0, 1, 2, \ldots \tag{7.12}$$

Any deviation from a circular interference pattern indicates a nonspherical surface. The spacing of the rings can be used to approximate the radius of curvature of the surface. Precise evaluation of the surface curvature using Newton's rings is best done with an accurate test plate that closely mates with the surface under examination. This has the effect of widely spacing the dark rings, allowing for a much more accurate evaluation of the surface radius.

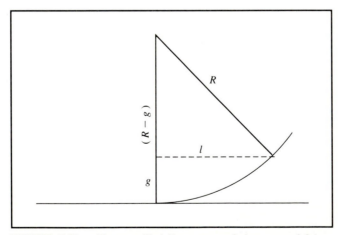

FIGURE 7.11. The sagittal theorem relates g and l for a spherical surface.

EXAMPLE 7.3	A convex lens surface has a radius of 14.82 cm. At what radius from the center will one find the first three dark rings when the surface is tested against an accurate flat using Newton's method with a source at 550 nm? The surface is then tested against an accurate concave surface of radius 15.00 cm. Where will one find the first dark ring then?

Solution The problem requires that the radii at which the rings are found be determined. Noting that

$$g = \frac{m\lambda}{2} \qquad m = 1, 2, 3, \ldots$$

in this case, one takes (7.11) as

$$l^2 = 2Rg - g^2$$

Since $g \ll R$, one typically uses the approximation

$$l = \sqrt{2Rg}$$

Here

$$l = \sqrt{14.82 \times 550 \times 10^{-7} \times m} \qquad m = 1, 2, 3, \ldots$$

from which $l = 0.285$, 0.405, and 0.494 mm and the dark rings are strongly clustered.

With the concave test plate one needs to find the value of l where the difference between the sagittal depths of the two plates is $\lambda/2$ so that

$$\frac{l^2}{2R_L} - \frac{l^2}{2R_T} = \frac{\lambda}{2}$$

where R_L is the radius of the lens and R_T that of the test plate. Solving for l one gets

$$l = \sqrt{\frac{\lambda R_L R_T}{R_T - R_L}}$$

and $l = 2.606$ mm. The advantage of the test plate is clear in that it widens the interference fringe pattern.

It is comparatively simple to illustrate the fringes found with Newton's rings, testing in one dimension using two flat plates such as microscope slides. If one sets the plates as shown in Figure 7.12, fringes will be visible as parallel bars across the plates. The fringes formed in this way are called *fringes of equal thickness* since the fringes index the

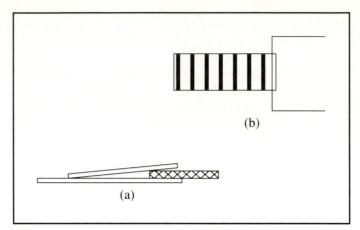

FIGURE 7.12.   Fringes of equal thickness. (a) A wedge formed using two microscope slides. (b) The fringe pattern observed.

thickness of the film at their level. The straight-line fringes formed in the Michelson interferometer when the mirrors are tilted relative to each other are similar to these since they index the relative wedge due to the tilt.

EXAMPLE 7.4

The arrangement in Figure 7.12 is to be used to determine the thickness of the paper used to form the wedge. The distance from the contact point of the slides to the edge of the paper is 50 mm. Monochromatic light at 628 nm is the source. There are 2.43 dark fringes visible per millimeter.

Solution

Each dark fringe corresponds to an additional separation of $\lambda/2$ since the path through the wedge is doubled. Thus, per millimeter,

$$\Delta t = \frac{\lambda}{2} \times 2.43 = 0.763 \quad \text{nm}$$

The wedge angle then is given by

$$\theta = \frac{0.763 \ \mu\text{m}}{1000 \ \mu\text{m}} = 0.000736 \quad \text{rad}$$

which is the tangent of the wedge angle, since for small angles the tangent is equal to the angle in radians. The thickness t of the paper is then given by

$$t = 50 \times 0.000763 \ \text{mm} = 0.038 \quad \text{mm}$$

This experiment is best done in monochromatic light since the colored fringes arising in white light are difficult to interpret. The fringes of equal thickness from white light are not uncommon. They are seen in oil films on wet road surfaces and also as a result of the separation of the layers of laminated windows in some older autos.

Antireflective Coatings

The fringes of equal thickness are visible both in transmission and in reflection. Those seen in an oil slick are reflective, while those observed through the separation of laminates are usually seen as transmitted fringes. A little thought will show that they are interrelated; that is, a reflective minimum will be a transmissive maximum as a result of energy conservation. This principle is used in "antireflective" coatings for lenses and other optical devices.

Consider an air–glass interface where the index of refraction of the glass is n_g. The *reflection coefficient,* the ratio of the incident to reflected intensity, is given by

$$R = \left(\frac{n_m - n_i}{n_m + n_i}\right)^2 \tag{7.13}$$

at normal incidence. For a crown glass with $n_m = 1.5$ and incidence from air $n_i = 1.00$, this gives 0.04, so that 4% of the incident light is reflected. Consider now a thin film of a material that is deposited on the surface with an index between that of air and of glass. Materials such as MgF ($n = 1.38$) and cryolite ($n = 1.36$) are appropriate, although these coatings work best with higher-index base materials. If the thickness of the deposited film is an odd number of quarter wavelengths, that is, if t in Figure 7.13 is given by

$$t = \frac{\lambda}{4}(2m + 1) \qquad m = 0, 1, 2, 3, \ldots \tag{7.14}$$

then the reflected light from the second surface of the film will be 180° out of phase with the incident light, and the reflected light will be an interference minimum, that is, zero. Note that λ in (7.14) is the wavelength in the medium, not the free-space wavelength; to use the free-space wavelength value λ_0 one must modify (7.14) and

$$t = \frac{\lambda_0}{4n}(2m + 1) \qquad m = 0, 1, 2, 3, \ldots \tag{7.15}$$

The best results arise if the ratio n_m/n_i is the same at each interface since that makes the reflection coefficient the same at each surface and the intensity change at the first surface is typically small.

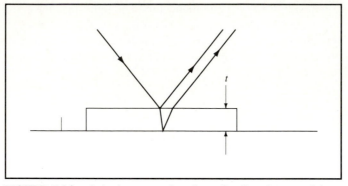

FIGURE 7.13.   Interference due to reflection from a thin film.

EXAMPLE
7.5

An antireflective coating is needed for a flint glass with index 1.695. The coating is to be of minimal thickness and function at 550 nm. Specify the coating.

Solution   A good antireflective coating requires that R at the first and second surfaces be approximately equal. That is,

$$\left(\frac{n_c - 1.0}{n_c + 1.0}\right)^2 = \left(\frac{n_g - n_c}{n_g + n_c}\right)^2$$

which gives

$$n_c^2 - n_g = 0$$

or

$$n_c^2 - 1.695 = 0$$

and $n_c = 1.302$. In this case the material of choice will be cryolite with $n = 1.360$. The film thickness for the quarter-wave coating is given by (7.15):

$$t = \frac{550}{4 \times 1.360}(2m + 1) \qquad m = 0, 1, 2, 3, \ldots$$

so that $t = 101.1, 303.3,$ and 505.5 nm for the first three possible coating thicknesses.

The deposition of the coatings in a nonreflective coating process is monitored interferometrically. To control the deposition process, which is done in a vacuum with the coating material being evaporated onto the surface, the practical procedure is to monitor the reflected light and stop the deposition when the reflected light goes to a minimum. It is important to recognize that, while the reflected light at a given wavelength is minimized, other wavelengths will still be reflected. In particular, minimizing the reflectivity at 550 nm, the wavelength where the eye is most sensitive, leads to enhanced reflectivity at the far red and far blue ends of the visible spectrum and a characteristic bluish-red cast to the surface of such antireflective-coated optics.

Interference Filters

While strongly monochromatic sources are available using lasers, often monochromatic sources are needed at wavelengths not covered by laser sources or for purposes where the laser would be too costly. One can then use a filter to get the required light. The simplest of these filters are dyed gelatin sheets, sometimes sandwiched between glass plates, called *Wratten filters*. These filters have rather broad passbands.

Much narrower passbands can be gotten using interference filters. These filters, called MDM (metal-dielectric-metal) filters, are composed of a dielectric coated on each side with very thin, partially transmissive metal films. The outer surfaces may be antireflective coated as well. Figure 7.14 illustrates the geometry with light incident from the upper left. The phase difference between two successive beams is $2\pi(2tn/\lambda)$, where t is the film thickness, λ is the wavelength in the dielectric, and normal incidence is assumed.

If the incident light is given by

$$a_i e^{j(\omega t - kx)} = a_i e^{j\theta} \tag{7.16}$$

and the amplitude reflection and transmission coefficients are ρ and σ, respectively,[2] the light transmitted through the filter is given by the sum of the successive waves exiting from the filter following each successive incidence on the second surface of the filter:

$$\sigma a_i e^{j\theta} + \sigma\rho e^{j(\theta - \phi)} + \rho^2 \sigma a_i e^{j(\theta - 2\phi)} + \cdots + \rho^n \sigma a_i e^{j(\theta - n\phi)} + \cdots \tag{7.17}$$

This can be rewritten

$$A = a_i e^{j\theta}(\sigma + \rho\sigma e^{-j\phi} + \rho^2 \sigma e^{-2j\phi} + \cdots) \tag{7.18}$$

which can be put in the form

$$A = \frac{a_i e^{j\theta}\sigma}{1 - \rho^{-j\phi}} \tag{7.19}$$

---

[2] These are the ratios of the incident and reflected *amplitude* and the incident and transmitted *amplitude*.

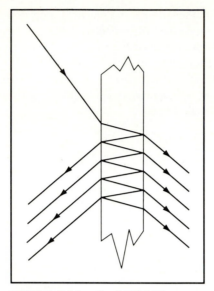

FIGURE 7.14.   An MDM filter.

The transmitted intensity I is given by

$$I = AA^* = \frac{a_i^2 \sigma^2}{(1 - \rho)^2} \frac{1}{1 + [4\rho/(1 - \rho^2)] \sin^2 \phi/2} \qquad (7.20)$$

The maximum intensity I_{max} is simply

$$I_{max} = \frac{a_i^2 \sigma^2}{1 - \rho^2} \qquad (7.21)$$

and the change in intensity with ϕ can be found from

$$\frac{I}{I_{max}} = \frac{1}{1 + [4\rho/(1 - \rho)^2] \sin^2 \phi/2} \qquad (7.22)$$

Maximum intensity occurs whenever $\phi = 2\pi m$, with m a nonzero integer. The bandwidth for the passband is defined as the band between the half-power points, that is, between the points where I/I_{max} has the value 1/2. If twice the half-width is given by α, then $\phi = 2\pi m + \alpha/2$ at the half-power level. If the coefficient of the sine term, called the *finesse*,

$$\mathscr{F} = \frac{4\rho}{(1 - \rho)^2} \qquad (7.23)$$

is large, the fringes are narrow and ϕ is small. This can be seen by expanding the sine term and taking the first term

$$\frac{1}{2} = \frac{1}{1 + \mathcal{F}(\alpha/4)^2} \tag{7.24}$$

so that

$$\alpha = \frac{4}{\sqrt{\mathcal{F}}} \tag{7.25}$$

If the reflection coefficient ρ is 0.5, corresponding to half-silvered mirrors for the metal surfaces, $\mathcal{F} = 8$, but with a ρ of 0.85, as is more typical of these filters, $\mathcal{F} = 151$ and the passband is narrow. Bandwidths at half-height as low as 20 nm are available as off-the-shelf items.

The mirrors that act as endplate mirrors for lasers are made of multiple layers of quarter-wave plates alternating between high and low index designed to maximize the reflectivity. With 15–20 layers, one gets reflectivities in excess of 99.9% of the incident light at the desired wavelength, that is, at the laser output wavelength.

PROBLEMS

7.1. In a Rayleigh interferometer the two chambers are filled with water. In chamber 1 the water is distilled, but in chamber 2 the water contains a dissolved salt. The illumination is with green 550-nm light, the arms are 30 cm long, and a shift of 6 fringes is noted. What is the difference in the index between the two liquids?

7.2. Would one expect to be able to use the Rayleigh interferometer to measure the absolute index of a liquid such as water? Suggest a method by which this interferometer could be used to find the index of liquids.

7.3. Prepare a phasor diagram for the four-slit interferometer and sketch the interference pattern.

7.4. A diffraction grating is 5 cm long and has 5000 slits. Find the angular separation between 400-nm and 600-nm radiation in both first and second order.

7.5. To measure the slit spacing in a grating, a laser operating at 598 nm is used. The first order appears at $\pm 47.3°$ from the central maximum. How many slits per centimeter are present in the grating?

7.6. The mercury spectrum is sometimes used as a standard in the measurement of spectra. The blue line is at 435.8 nm and the yellow line is at 579.1 nm. The grating in the spectrometer has 475 slits per millimeter. In what order must the system be used if one requires that these lines be separated by at least 5°?

7.7. The minimum separation of two spectral lines that can be resolved by a grating $\Delta\lambda$ is

$$\Delta\lambda = \frac{\lambda}{Nm}$$

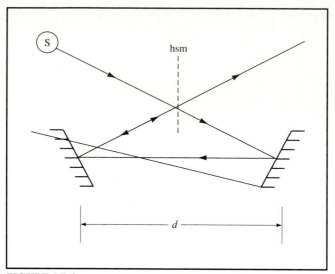

FIGURE P7.9.

where N is the total number of slits in the grating and m is the order. What is the minimum number of slits per centimeter in a 3-cm grating that will resolve the sodium D lines at 589.0 nm and 589.6 nm in third order?

7.8. How many slits would a grating need to separate in third order the red emission lines in hydrogen and deuterium which occur at 656.6 nm and are separated by 0.18 nm?

7.9. A triangular path interferometer has the form shown in Figure P7.9. This is to be used with 1-cm microwaves. Discuss the setup, particularly the reflector separation d.

7.10. A Sagnac interferometer 8 cm on a side has a 0.003-fringe shift with 550-nm light. What is the angular velocity of the interferometer about its center of rotation?

7.11. Two radio antennas operate at 840 kHz and are spaced 125 m apart. If the antennas oscillate in phase, how many maxima will there be in a circle centered between them and 10 km from them?

7.12. In order to observe Newton's rings in a planoconcave lens, a planoconvex lens with $R = 80.01$ cm is placed atop it. Using 550-nm light, the 10th dark ring is found at a radius of 12 mm. What is the radius of curvature of the concave surface?

7.13. A soap bubble 0.5 μm thick is viewed in white light at normal incidence. What color is transmitted?

7.14. An antireflective coating of MgF ($n = 1.375$) is applied to a glass lens. How thick must it be to produce minimum reflection at 550 nm? Will any light at 550 nm be reflected?

7.15. What would be the optimal index for the lens in Problem 7.14 with a MgF coating?

7.16. Cryolite ($n = 1.360$) is to be used to make an MDM filter for 476-nm light. How thick should it be? What is the effect of tilting the filter? If one utilized tilt, what might be the range over which this filter could be used?

7.17. The surface reflectivities of the MDM filter in Problem 7.16 are 0.72. What are the finesse and the bandwidth at the half-power points?

CHAPTER 8

Diffraction

A detailed examination of the edge of a shadow shows that there is a structure to this edge; it is not the sharp, stepped edge one would expect based on the geometrical model. In this chapter the effect of selecting a portion of a wavefront by imposing an opaque object is examined. Figure 8.1 shows the amplitude distribution in the geometrical shadow of a straightedge. The light distribution from an isolated portion of a wavefront will be derived. The results that will be found here are of great importance in many optical phenomena ranging from optical signal processing to metrology.

As the wavefront is truncated in passing through an aperture in a screen, its amplitude distribution begins to be altered almost immediately. The ''pulse-shaped'' wave quickly loses its plane front and begins to develop a structure. The structure is far from static, as can be seen in Figure 8.2, where the shape of the wavefront is shown as its distance from the aperture increases. Once past the aperture it is governed by the Kirchhoff–Fresnel equation (AII.25) provided that it is a large number of wavelengths beyond the slit.[1] The region to the right of the slit in Figure 8.2 is normally broken down into two separate regions that are related to the structure of the wavefront as well as to the approximations made in reducing it (AII.25) to a form that lends itself to solution.

The two regions that are usually examined are the *near-field or Fresnel region* and the *far-field or Fraunhofer region*. The transition from Fresnel region to Fraunhofer region as one moves away from the aperture is rather vague. The Fresnel solution is valid from a region near the slit out to infinity, while the Fraunhofer solution is valid when one is well removed from the aperture where both approximations lead to the same result.

---

[1] The development of the equations governing the light disturbance for a truncated wavefront is given in Appendix II.

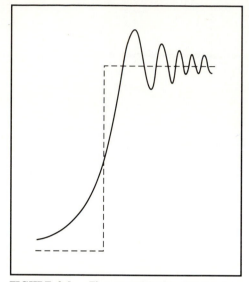

FIGURE 8.1.   The amplitude distribution at the edge of the geometrical shadow of a straightedge. Note that at the geometrical edge of the shadow, the amplitude is only one-half that which would be expected based on a geometrical model.

Fresnel Diffraction

The near-field or Fresnel diffraction region begins at a distance many times the aperture dimension beyond the slit, and the Fresnel solution is valid in a region a few tens of degrees about the direction of propagation of the wave U_1 incident on the aperture Σ as in Figure 8.3. Throughout this discussion U_1 will be assumed to be a plane wave so that both the amplitude across Σ and the relative phase will be constant.

Initially, to evaluate the optical disturbance at $U(x_0, y_0, z)$ in the diffraction field, Huygens's principle will be used. Huygens developed the idea that points on a wavefront serve as secondary sources and that their combination at some distant point fixes the optical disturbance there. Figure 8.4 shows how successive wavefronts are constructed from Huygens's principle. The formal development of Huygens's principle has basis in Kirchhoff's theorem as given in Appendix II.

One can begin by examining the intensity distribution near the edge of the shadow of a straightedge illuminated by a plane wave as shown in Figure 8.1. The geometrical model leads to the conclusion that the intensity is constant in the region of illumination and zero elsewhere. In the wave approximation the result is somewhat different. The point P_0 in Figure 8.5 is just at the edge of the geometrical shadow. The region above the

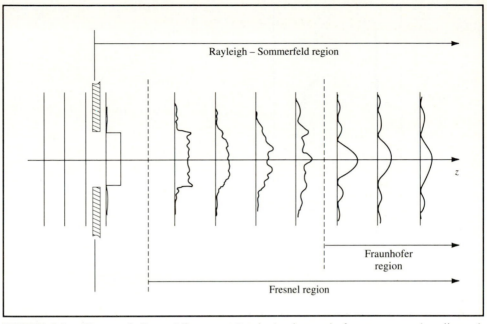

FIGURE 8.2. The evolution of the wavefront of a truncated wave as a function of the distance from the aperture. The regions denoted Fresnel and Fraunhofer refer to the form of the solution to the diffraction equations in that region.

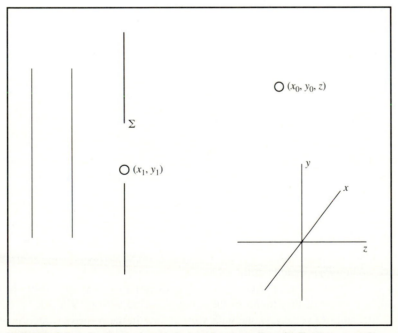

FIGURE 8.3. A plane wave is incident on the slit Σ. The coordinates in the slit are (x_1, y_1) and in the field (x_0, y_0, z).

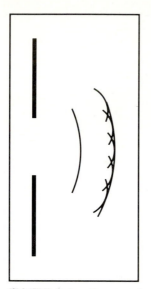

FIGURE 8.4. Huygens's
principle.

straightedge contributes to the illumination at P_0, while nothing passes the straightedge
screen Σ. The region above Σ may be thought of as divided into zones where the incremental change in path length between successive zones is one-half wavelength. The
illumination at P then is the sum of the contributions of all the zones lying above Σ. There
are two factors besides the amplitude of the wave that affect the illumination at P_0. Since
we are treating this problem in terms of Huygens's principle, the waves from each point
on the wavefront arising at the position of the screen are spherical waves and thus have a
$1/r$ amplitude dependence. In addition, as can be seen in (AII.28), there is an *obliquity
factor;* the consequence of obliquity is that the amplitude of the wave received at P_0 from
the zone y_n in Figure 8.5 depends on $\cos(\mathbf{n},\mathbf{r}_{yn})$, where $(\mathbf{n},\mathbf{r}_{yn})$ is the angle between the
normal to the wavefront and $\mathbf{r}_{yn}$.

   The optical disturbance at P_0 can be viewed as an infinite series of terms. The
arbitrary division above the screen with successive zones differing by one-half wavelength
allows one to see that if one treats the zones pairwise, the second zone nearly cancels the
first. The cancellation is not complete since both the $1/r$ terms and the obliquity factor act
to make the magnitude of the contribution of each successive zone somewhat smaller. The
infinite series of terms arising from this summation of zones does converge, however, and
the limiting term is zero.

   Qualitatively, one can move P_0 up and down vertically and consider the disturbance. When P_0 is exactly at the edge of the geometrical shadow, each successive pair of
terms gives a positive contribution to the disturbance and the result is a disturbance
amplitude one-half that found in the absence of the screen. With no screen, one would
have the sum of two series, one with y increasing in the positive direction and one with y
increasing in the negative direction. The intensity, the square of the disturbance, at the

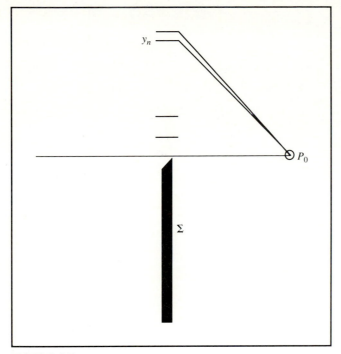

FIGURE 8.5.

edge of the screen is one-fourth that in the absence of the screen. As P_0 moves in a negative y direction below the edge of the screen, less and less of the first positive zone contributes to the disturbance and the disturbance falls to zero.

As P_0 moves in a positive y direction, it reaches a maximum when it is situated so that the edge of the screen falls at the edge of the first zone. As it moves farther and farther above the edge, the disturbance, and thus the intensity, oscillates with smaller and smaller amplitude until the contribution of each additional pair of zones becomes vanishingly small.

EXAMPLE 8.1

A shadow is formed on a screen 1 m behind a straight-edged obstruction (Figure E8.1). Treat the light as a monochromatic plane wave with wavelength 500 nm. How wide is the first "zone" as defined earlier? What is the relative amplitude of the tenth zone relative to the first? What is the ratio of the disturbance of the tenth to the eleventh zone?

Solution

The width of the first zone δ is given by

$$\delta^2 = \left(r + \frac{\lambda}{2}\right)^2 - r^2$$

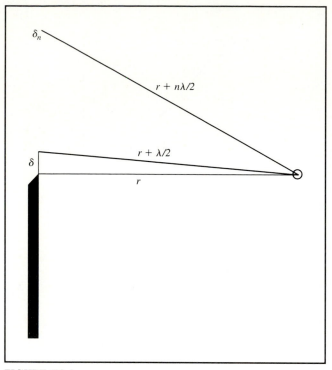

FIGURE E8.1.

or

$$\delta \simeq \sqrt{r\lambda}$$

since

$$r\lambda \gg \frac{\lambda^2}{4}$$

Here

$$\delta \simeq \sqrt{5 \times 10^{-4} \times 1 \text{ m}^2}$$

that is, 7.1×10^{-4} m or 0.71 mm.

This will be the height above the edge of the geometrical shadow at which one will find the maximum intensity.

δ_{10} is found where the path length is $r + 5\lambda$ or at a height of

$$\delta_{10} \simeq \sqrt{5\lambda r} = \sqrt{5 \times 5 \times 10^{-7} \text{ m} \times 1 \text{ m}} = 15.8 \times 10^{-4} \quad \text{m}$$

or 1.58 mm.

Relative amplitude may be found by taking the nth zonal contribution (AII.29) as

$$A_0 \frac{\cos(\mathbf{n},\mathbf{r})}{[r + n(\lambda/2)]} = A_0 \frac{r}{[r + n(\lambda/2)]^2}$$

The contribution of the first zone is

$$A_0 = \frac{r}{(r + \lambda/2)^2}$$

The tenth zone compared to the first gives

$$\frac{A_{10}}{A_1} = \left(\frac{r + \lambda/2}{r + 5\lambda}\right)^2 = \left(\frac{1.00000025}{1.0000025}\right)^2 = 0.99999550$$

while

$$\frac{A_{10}}{A_{11}} = 0.99999950$$

The small difference between successive terms suggests that the series converges slowly, and that is the reason the structure of a shadow's edge is not readily apparent.

The concept of the ''Fresnel zone'' is perhaps best illustrated by the circular plate shown in Figure 8.6, known as a *Fresnel zone plate*. Here a circular aperture in a screen is covered by a transparent plate in which alternate zones are darkened. This case, in contrast to that of a straightedge, has the ''canceling'' zone removed, and all the zones viewed from the appropriate position behind the screen are additive.

One can design a zone plate for a given distance by establishing the radii of successive zones at the design distance. The geometry is illustrated in Figure 8.7. The point of interest, O, is at a distance R from the screen Σ, and one needs to find r_n such that each increment in r_n increases the path length to O by $\lambda/2$. Once the zonal radii are found, the alternate zones are made opaque and then one has only positive contributions to the optical disturbance at O. Using the Pythagorean theorem for the triangle with legs R and r_n, one has

$$R^2 + r_n^2 = (R + n\lambda)^2 \tag{8.1}$$

so that

$$r_n = \sqrt{2n\lambda R + n^2\lambda^2} \tag{8.2}$$

Since $\lambda \ll R$, one usually drops the last term in the square root above and uses

$$r_n \simeq \sqrt{2nR\lambda} \tag{8.3}$$

FIGURE 8.6.   The Fresnel zone plate.

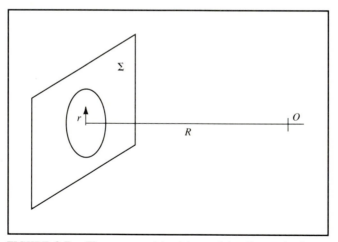

FIGURE 8.7.   The geometrical layout for the calculation of Fresnel zones.

**EXAMPLE
8.2**   Calculate the first three zonal radii for a point on the optic axis 50 cm from a screen to be used as a zone plate illuminated with 550-nm light. If one then uses these values to form a three-zone zone plate, what will happen to the intensity along the axis of the system as one approaches the screen from the 50-cm position?

Solution   The first three zones can be found using equation (8.2) with $n = 1, 2, 3$:

$$r_n = \sqrt{2 \times n \times 550 \times 10^{-7} \text{ cm} \times 50 \text{ cm}}$$

or

$$r_n = \sqrt{5.5 \times 10^{-2} \times n} \text{ cm}$$

so that $r_1 = 0.2345$ cm, $r_2 = 0.3317$ cm, $r_3 = 0.4062$ cm, and r_3 gives the radius of the hole in the screen needed to accommodate the zone plate.

As one moves along the optic axis toward the screen each zone now subtends more than a single zone and the intensity begins to decrease. When one reaches the point on the axis where r_1 represents the radius of two zones, that is,

$$r_1^2 = (0.2345)^2 \text{ cm}^2 = 2 \times 1 \times 5.5 \times 10^{-5} \text{ cm} \times R \text{ cm}$$

or $R = 45.45$ cm, the intensity will reach a minimum. This minimum will not be zero because of obliquity factors, but it will be a clear minimum. As one reaches the point on the axis where 0.2345 cm represents precisely three zones, one will have a maximum that will be followed by another minimum, so one finds that the intensity is modulated along the optic axis.

As can be seen from the example, the zone plate can be used to focus light, and therefore it can function as a lens. This "lens" forms multiple images so that it is less efficient than a lens that concentrates all the energy collected by it into a single image. In addition, the fidelity of the image degrades much more rapidly as one moves away from the optic axis than is the case with the typical lens.

The Fresnel Integral

The optical disturbance at points in the Fresnel region can be found using the Fresnel integrals (AII.42) which are derived in Appendix II.

$$\int_{u_L}^{u_U} \cos \frac{\pi}{2} (u')^2 \, du' = C(u') \qquad \int_{u_L}^{u_U} \sin \frac{\pi}{2} (u')^2 \, du' = S(u') \qquad (8.4)$$

u_L and u_U are variables related to the lower and upper edges of the aperture by

$$u = \frac{x_0 - x_1}{\sqrt{\lambda Z}} \qquad (8.5)$$

TABLE 8.1   Fresnel Integrals[a]

u	C(u)	S(u)	u	C(u)	S(u)
0.0	0.0000	0.0000	3.2	0.4663	0.5934
0.2	0.1999	0.0042	3.4	0.4385	0.4296
0.4	0.3975	0.0334	3.6	0.5880	0.4923
0.6	0.5811	0.1105	3.8	0.4481	0.5656
0.8	0.7228	0.2493	4.0	0.4984	0.4205
1.0	0.7799	0.4383	4.2	0.5417	0.5632
1.2	0.7154	0.6234	4.4	0.4383	0.4623
1.4	0.5431	0.7135	4.6	0.5672	0.5162
1.6	0.3655	0.6389	4.8	0.4338	0.4968
1.8	0.3336	0.4509	5.0	0.5636	0.4992
2.0	0.4882	0.3434	5.2	0.4389	0.4696
2.2	0.6363	0.4557	5.4	0.5573	0.5140
2.4	0.5550	0.6197	5.6	0.4517	0.4700
2.6	0.3889	0.5500	5.8	0.5298	0.5461
2.8	0.4675	0.3915	6.0	0.4995	0.4469
3.0	0.6057	0.4963	∞	0.5000	0.5000

[a]$C(-u) = -C(u);\ S(-u) = -S(u)$.

where (x_0, y_0, z) is the position of the field point, $(x_1, y_1, 0)$ is the position within the aperture, and λ is the wavelength of the light. These integrals do not have a simple solution but their values are tabulated. Table 8.1 is a table of the Fresnel integrals. The amplitude at a point (x_0, y_0, z) is given by (AII.43),

$$U(x_0, y_0, z) = \frac{A}{2} jk e^{jkz} \left\{ \begin{array}{l} [C(x_U) - C(x_L)] - j[S(x_U) - S(x_L)] \\ \times\ [C(y_U) - C(y_L)] - j[S(y_U) - S(y_L)] \end{array} \right\} \qquad (8.6)$$

and the corresponding intensity is given by (AII.44),

$$I(x_0, y_0, z) = \frac{A^2}{4} \left\{ \begin{array}{l} [C(x_U) - C(x_L)]^2 + [S(x_U) - S(x_L)]^2 \\ \times\ [C(y_U) - C(y_L)]^2 + [S(x_U) - S(x_L)]^2 \end{array} \right\} \qquad (8.7)$$

EXAMPLE 8.3

A slit in a screen has width 0.4690 cm. Find the intensity at $(0,0,50)$ if the origin is at the center of the slit. The slit is illuminated with 550-nm light having unit amplitude from a remote source.

Solution

This is a one-dimensional problem since the limitation of the wave is only in one direction; that is, the slit is assumed to be infinitely long. The values of x_U and x_L are ± 0.2345 cm. At $(0,0,50)$ the values of u'_U and u'_L are given by

$$u' = \frac{\pm 0.2345}{\sqrt{550 \times 10^{-7} \times 50}}$$

and interpolating in Table 8.1 one finds $C(u_U) = 0.4845$, $S(u_U) = 0.4816$, $C(u_L) = -0.4845$, and $S(u_L) = -0.4816$. The intensity is given by (8.7) and

$$I(0,0,50) = \frac{1}{4}[(0.9690)^2 + (0.9632)^2] = \frac{1.8667}{4} = 0.4667$$

An alternative graphical procedure using the Fresnel integrals is useful in understanding Fresnel diffraction. Figure 8.8 is a graph representing $C(u)$ and $S(u)$ for all values of u. The curve is plotted with $C(u)$ as the abscissa and $S(u)$ as ordinate and is known as the *Cornu spiral*. For given values of u_U and u_L the length of the line joining these two points is proportional to the amplitude of the disturbance at the field position.

As an example, consider the amplitude distribution at the edge of the shadow of a straightedge as illustrated in Figure 8.1. The value of u_L' will be zero at the edge of the shadow, and u_U' will be infinity. At the edge the amplitude vector **a** in Figure 8.9 will extend from the origin, where u' is zero, to the center of the upper arm of the spiral, where u' is infinite. This is precisely one-half the distance from the center of the lower arm of the spiral, so the amplitude is simply one-half that found in the absence of the screen. As one moves above the axis, the value of u_L' will become negative and one will have amplitudes represented by **b, c, d,** and so on. The oscillatory behavior of the amplitude

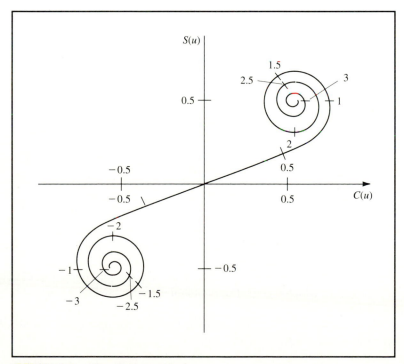

FIGURE 8.8. The Cornu spiral.

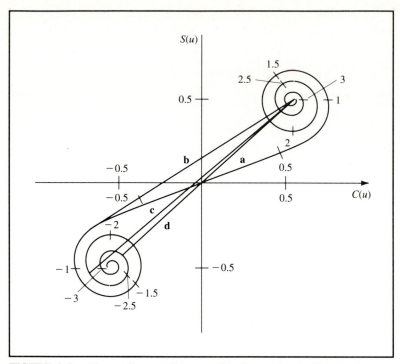

FIGURE 8.9.   The values of the amplitude vector in the region above the geometrical edge of the shadow of a straightedge.

will be clear. One the other hand, as one moves into the shadow, u'_L will track through successively larger positive values of u and the continued decrease in the amplitude of the disturbance will be clear from **b, c, d,** and so on, as shown in Figure 8.10. The Cornu spiral then can be seen to give a rapid qualitative understanding of the optical disturbance in the diffraction region.

The Fraunhofer Region

The Fraunhofer region is in the far field, many times farther from the sources than the Fresnel region. The simplest approach to examination of the far field is to evaluate the optical disturbance at infinity where the Fraunhofer approximation is clearly valid. That approach will be adopted here. The Fraunhofer approximation is derived from the Fresnel approximation.

To examine the Fraunhofer region one first compares the kernel of the Fresnel diffraction equation (AII.36), namely,

$$K = U(x_1, y_1)e^{j(k/2z)(x_1^2 + y_1^2)} \tag{8.8}$$

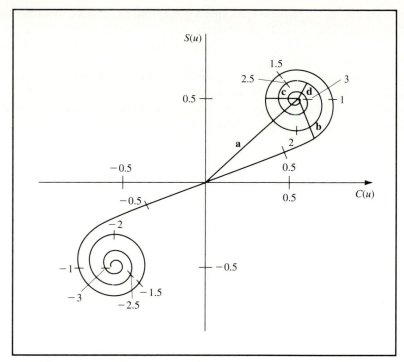

FIGURE 8.10. The variation of the amplitude vector of the optical disturbance within the region of the shadow of a straight-edge.

with that found in the Fraunhofer diffraction equation (AII.40),

$$U(x_0,y_0) = \frac{Ae^{jkz}}{j\lambda z} \int_{-\infty}^{+\infty}\!\!\int U(x_1,y_1)e^{(2\pi j/\lambda z)(x_0 x_1 + y_0 y_1)}\, dx_1\, dy_1 \qquad (8.9)$$

where the kernel of the Fraunhofer integral can be seen to be simply the optical disturbance in the aperture of the screen $U(x_1,y_1)$. The difference between the two kernels of (8.8) and (8.9) is a term dependent upon the phase of the disturbance within the aperture that is present in the Fresnel kernel and not in the Fraunhofer kernel. From this one can understand the Fraunhofer approximation as one in which the phase differences across the aperture in the plane of the screen are negligible.

One might ask, for a slit with an aperture of 1 mm, how far would one need to be for the phase change across the slit to be minimal, say less than π radians. This would mean, using (8.8), that

$$\frac{k}{2z}(1)^2 = \pi \qquad (8.10)$$

and with light at 500 nm, z will be of the order of 2×10^6 m.

At infinity all radiation reaching a point from an aperture in a screen is parallel, as is shown in Figure 8.11. To facilitate the observation, one usually uses a lens to gather the light, which then maps the diffraction figure from infinity onto the focal plane of the lens as shown in Figure 8.12.

A qualitative understanding of what happens in the Fraunhofer region can be developed in the following way: On the axis of the slit where θ in Figure 8.12 is zero, the total disturbance reaching the observing plane can be thought of as made up of the sum of the contributions of many small regions in the slit. These contributions are in phase, since in the Fraunhofer approximation there is no phase change from the radiation across the slit. In Figure 8.13 all these small contributions lie along the abscissa for the axial point, $x_0 = 0$, $y_0 = 0$, $z = \infty$. Their sum A' is the amplitude of the on-axis disturbance at infinity.

As one moves from the axis to angle θ in Figure 8.11, the contributions of the individual subregions remain the same in magnitude, but they now have a phase difference between them arising from the path change within the diffraction region. Their resultant in Figure 8.13 is the observed amplitude at angle θ. In the limit, as the width of the individual subregions goes to zero, one has the situation illustrated in Figure 8.14. The arc C is the limiting case of an equilateral polygon as the number of sides of the polygon increases without bound. The arc is a circular segment. The length of C is the scalar amplitude at an angle that is equal to $|A|$ if one ignores any obliquity factors. The angle θ in Figure 8.14 is the phase difference between the lower and upper margins of the slit. It is also the angle subtended by C. The length of C, the scalar amplitude at angle θ, is

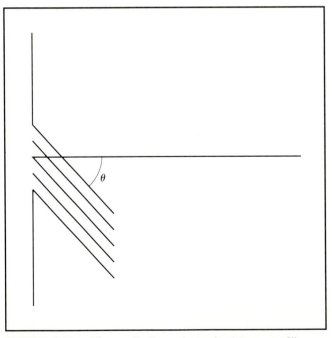

FIGURE 8.11.   The radiation at angle θ from a slit.

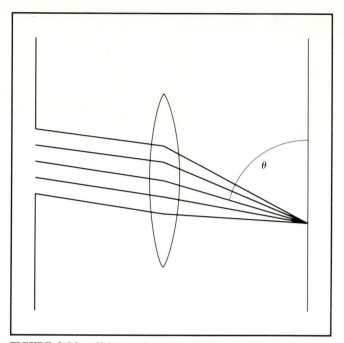

FIGURE 8.12. Using a lens to display the Fraunhofer diffraction figure.

simply $R\theta$, which is the same as $|A|$. The vector amplitude A is the chord of the arc C, and it is given by

$$A = 2R \sin(\theta/2) \tag{8.11}$$

The ratio of the vector to the scalar amplitude G is

$$G = \frac{\sin(\theta/2)}{\theta/2} \tag{8.12}$$

and the amplitude as a function of θ varies as shown. The intensity varies as G^2 or

$$I_{\text{obs}} = \frac{\sin^2(\theta/2)}{(\theta/2)^2} \tag{8.13}$$

where I_{obs} is the intensity of the on-axis point where $\theta = 0$.

Figure 8.15 shows the intensity distribution. Even though this is a qualitative argument, the result is reasonably accurate. In practice there is an obliquity term, but its effect is small. Equation (8.13) then gives the intensity dependence but not the amplitude term, which must be gotten from the quantitative treatment of the problem.

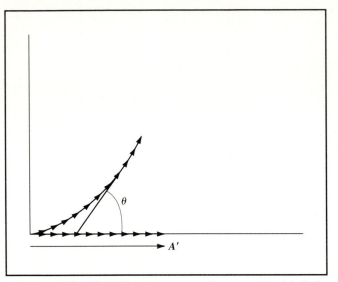

FIGURE 8.13.   The disturbance on the screen at infinity on-axis and at an angle θ.

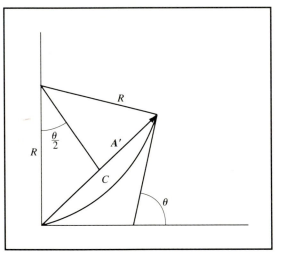

FIGURE 8.14.

If θ is large enough, say, 4π, then the arc of Figure 8.14 becomes a full circle as in Figure 8.16 and the vector A' goes to zero. As a result I is also zero at that angle. The disturbance vanishes each time the angle increases by a multiple of 4π. The intermediate maxima between these null points are successively smaller due to the denominator of $(\theta/2)^2$.

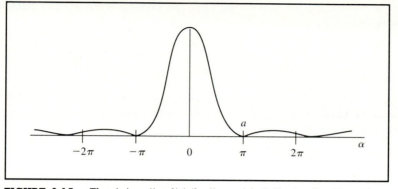

FIGURE 8.15. The intensity distribution at infinity for the Fraunhofer diffraction pattern of a slit.

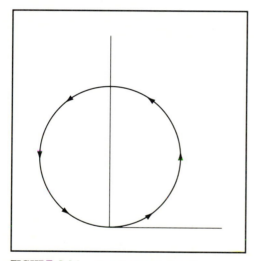

FIGURE 8.16. $\theta = 2\pi$ and $A = 0$.

Equation (8.9) can be solved quantitatively using the substitution (AII.37):

$$f_x = \frac{x_0}{\lambda z} \qquad f_y = \frac{y_0}{\lambda z} \tag{8.14}$$

For a slit of width 2ξ parallel to the x axis, one has a one-dimensional problem and (8.9) takes the form

$$U(y_0) = \frac{e^{jkz}}{jkz} e^{j(k/2z)y_0^2} \int_{-\xi/2}^{+\xi/2} A e^{j\,2\pi f_y y_1}\, dy_1 \tag{8.15}$$

for a plane wave incident on the slit. The integral is the Fourier transformation for a rectangular pulse of width ξ and amplitude A^2. This gives for $U(y_0)$

$$U(y_0) = \frac{e^{jkz}}{jkz} e^{j(k/2z)y_0^2} A \frac{\sin \pi \ (\xi/2f_y)}{\pi \ (\xi/2f_y)} \tag{8.16}$$

which can be seen to have the same form as equation (8.13).[2] The coefficient is somewhat different, but the dependence on $(\sin \theta)/\theta$ is the same so that the diffraction figure has the shape shown in Figure 8.15.

EXAMPLE 8.4

Two very narrow slits separated by a distance $d = 0.1$ mm are cut into an opaque screen. The screen is illuminated with light at 550 nm, and a converging lens immediately behind the screen brings the light to a focus. What does the diffraction figure look like?

Solution

This is a one-dimensional problem as earlier. What is required is the Fourier transform of the two slits. The function $U(y_1)$ is given by

$$U(y_1) = A \left[\delta\left(y_1 - \frac{d}{2}\right) + \delta\left(y_1 + \frac{d}{2}\right) \right]$$

where $\delta(y)$ is the delta function defined in Appendix II and its Fourier transform can be easily evaluated and shown to be equal to

$$F(f_y) = 2A \cos 2\pi f_y \frac{d}{2}$$

The amplitude in the observing plane then is a cosinusoidal function. This is precisely Young's experiment. The intensity of the diffraction figure is proportional to a cosine-squared term. Intensity maxima occur with squared sinusoids whenever the argument is $n\pi$, n an integer. In this case

$$\pi f_y \frac{d}{2} = n\pi$$

Substituting for $n\pi$ and noting that y_0/z is the angle θ, one gets

$$\theta = \frac{n\lambda}{d} \qquad n = 0, 1, 2, 3, \ldots$$

This can be compared with the result gotten in Chapter 6 for Young's experiment.

The result in equation (8.16) forms the basis for the understanding of the resolution of an optical system. Consider two plane waves arising from two line sources at infinity

---

[2] A review of the Fourier transform appears in Appendix III.

which generate the plane waves which fall on the slit with their wave vectors separated by a small angle ψ as in Figure 8.17. Each of these waves generates a diffraction figure like that shown in Figure 8.15. The arbitrary definition of resolvability put forward by Rayleigh is that the maximum of one figure be no closer to the second than its first minimum, marked as a in Figure 8.15. The angular separation of the waves between the central maximum and the first minimum, that is, where

$$\pi f_y \frac{\xi}{2} = \pi \tag{8.17}$$

when f_y is replaced with (8.14), yields

$$y_0 = \frac{\lambda z}{\xi/2} \tag{8.18}$$

Equation (8.19) comes about from a one-dimensional slit problem. Most often the question of resolution limit arises with two point sources imaged through a lens with a circular aperture. This then requires use of (8.9) and a two-dimensional Fourier transformation. As indicated in Appendix III, when a two-dimensional circular pulse is transformed, the solution is in the form of a Bessel function. The Bessel function has a form similar to $(\sin\theta)/\theta$, as one would intuitively expect. The singular change produced here is a change in the coefficient of the separation of the images so that

$$R = 1.22 \frac{\lambda f}{D} \tag{8.19}$$

where R again is the angular separation of the image points and D is the diameter of the aperture.

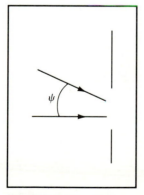

FIGURE 8.17. The wave vector directions for two closely spaced sources at infinity.

Babinet's Principle

The diffraction problems just treated have all been based on apertures in a screen such as that in Figure 8.18a. The *complementary screen*, Figure 8.18b, is one in which only the aperture of (a) is opaque. The diffraction figure for the aperture in Figure 18a can be gotten by solving the diffraction integral. What about the diffraction figure of the complementary screen? This too could be gotten by solving the diffraction integral, but a simpler method exists that invokes the conservation of energy. The plane wave falling on the screen has an amplitude of A over the entire screen, remembering that a plane wave is one in which the amplitude is the same everywhere in space.

The presence of the screen subtracts from the uniform intensity of the wave that portion of the wave falling on it. Only the energy passing through the aperture is available to form the diffraction figure, and that figure can be found as was seen earlier. If the amplitude distribution for the aperture in Figure 8.18a is taken as E_a and that of the plane wave as A, then the amplitude distribution from Figure 8.18b, the complementary screen, is

$$E_b = A - E_a \qquad (8.20)$$

which is *Babinet's principle*. This principle is most useful in the calculation of Fraunhofer diffraction patterns.

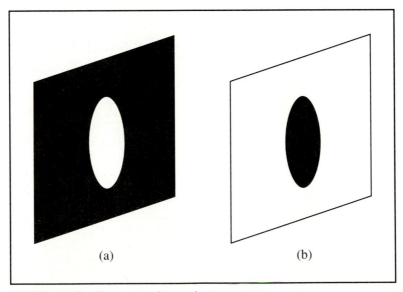

(a)   (b)

FIGURE 8.18.   Two complementary screens.

One important result is for the case of a collimated beam of light. The regions outside the beam have $A = 0$, and then

$$E_b^2 = (E_a)^2$$

which shows that the result of diffraction by a small hole or by a particle of the same diameter in the beam will produce the same effect at the points where $A = 0$ except for a $180°$ phase shift from the negative sign in (8.20). The intensity of the disturbance in these regions, proportional to the amplitude squared, is the same in either case.

PROBLEMS

8.1. Plane waves with 550-nm wavelength fall on a screen with a circular hole 4.44 mm in diameter. A viewing screen is placed 1 m from the hole. What does one find on the point lying along the axis of the screen?

8.2. A screen has an aperture with a 4-mm radius, and it is illuminated by plane waves falling normally on it. The waves have wavelengths of 550 nm and 632 nm. Where on the axis can one find monochromatic light at either wavelength?

8.3. Plane monochromatic light of wavelength 600 nm is incident on a 2-cm zone plate. The focus of the plate is at 80 cm. How many transparent zones are there in the plate?

8.4. Plane white light is normally incident on the zone plate described in the previous problem. Describe qualitatively what one would observe on the axis as the location of the viewing screen is varied.

8.5. The sixth boundary of a zone plate has a diameter of 4 mm. What is the principal focal length for 550-nm light, and what are the first two secondary focal lengths?

8.6. In the x-ray region of the spectrum, one cannot easily make a lens since most materials have an index of refraction close to 1. To focus 2-nm x rays, a zone plate with 25 zones is to be created. The diameter of the zone plate is 2 mm. What is its focal length?

8.7. Use the Cornu spiral in Figure 8.8 to find the Fresnel diffraction of a pair of equally wide slits separated by twice their widths.

8.8. A screen is pierced with a hole which is 1 cm in diameter. How far must one be from the screen so that the phase change across the hole, when viewed from a point on the hole's axis, is less than $\pi/3$ rad? Assume that the screen is illuminated with plane 550-nm light.

8.9. The headlights of an approaching car are 130 cm apart. Assume an average wavelength of 550 nm, where the eye is most sensitive, and estimate the distance at which the eye would resolve the two lights. Assume a 5-mm pupillary aperture for the eye.

8.10. A wire 100 μm in diameter lies in the path of a 550-nm plane wave. Use Cornu's spiral to find the diffraction pattern on a screen 1 m from the wire.

8.11. Compare the diffraction patterns observed for a hole in a screen 1 mm in diameter and the complementary 1-mm opaque object when placed at the same distance from the screen.

8.12. Lasers can be so well collimated that the beam spreading is due only to diffraction effects. The output of a He–Ne laser at 632.8 nm is 2.5 mm in diameter. What is the spot diameter at 500 m? 10,000 m?

8.13. A single slit is illuminated by two monochromatic plane waves and the diffraction pattern is viewed in the focal plane of a converging lens with a 1-m focal length. The slit is 400 μm wide, and the fourth minimum of the first wave and the fifth of the second wave coincide at 5 mm from the axis. What are the wavelengths?

8.14. The antenna pattern of a radar antenna can be gotten from the diffraction pattern of the antenna. Assume 3-cm microwaves and a rectangular antenna 1 m by 2 m. What is the angular size of the central lobe of the pattern?

CHAPTER 9

Light Sources—Lasers

Throughout this book little has been said about the source of the light which was being studied. In this chapter that omission will be rectified, with particular emphasis being placed on the laser, the most modern light source and the principal source of highly coherent light.

Most light sources from the sun to the street light to your desk lamp derive their light through a conversion of heat or chemical energy into electromagnetic radiation. Incandescent lamps emit light from an electrically heated filament, sodium vapor lamps from sodium gas in a tube excited by an electrical discharge, while the light emitted by a firefly in a garden comes from a cold, phosphorescent chemical interaction. Each of these cases involves a material medium which emerges as a necessary component of the production of light radiation.

Thermal Radiation

The radiation emitted by a hot body such as the sun or a lamp filament is continuous in wavelength. That is, the spectrum is complete and spans the visible range if the emitter is sufficiently hot. The distribution of radiation from a thermal source in equilibrium is known as the *Stefan–Boltzmann law*

$$I = \sigma T^4 \tag{9.1}$$

where σ is a constant, 5.67×10^{-8} W/m²-K⁴, so that an ideal radiator will radiate 56.7 W/cm² at 1000 K. Most radiation sources are not perfect, however, so that it is necessary to modify equation (9.1) by the introduction of a coefficient ϵ:

$$I = \epsilon \sigma T^4 \tag{9.2}$$

where ϵ varies between 0 and 1. A perfect radiator with ϵ equal to 1 is known as a *blackbody,* and, as a consequence of thermodynamics, it is also a perfect absorber. Those objects with ϵ less than 1 are sometimes called *gray bodies.*

The spectral distribution of a hot source, while continuous in both wavelength and intensity, has a temperature-dependent intensity distribution. The expression governing this distribution is known as *Planck's radiation law* and is written

$$I_\nu = \frac{2\pi h \nu^3}{c^2} \frac{1}{e^{h\nu/kT} - 1} \tag{9.3}$$

where h is *Planck's constant,* 6.67×10^{-34} J-s; c is the velocity of light; k is Boltzmann's constant, 1.38×10^{-23} J/K; and ν is the frequency of radiation. This law is consistent with the Stefan–Boltzmann law. The shape of the Planck radiation law is given in Figure 9.1, where λ is the abscissa and the intensity I_λ is the ordinate. Clearly the area under the curve must be finite as the Stefan–Boltzmann equation (9.1) indicates. Also since at $T = 0$ K the object will not radiate, I_λ must have a maximum value as may be seen in Figure 9.1. An object which is "red hot" is an object which produces most of its visible radiation in the red region of the spectrum. A device known as a pyrometer is used to measure the temperature of hot radiating objects by comparing the radiation spectrum of a wire heated to a known temperature with that of the hot object.

EXAMPLE 9.1

How does the frequency (or wavelength) of I_{max} vary with T?

Solution

This is most easily approached by making the substitution

$$x = \frac{h\nu}{kT}$$

in Planck's equation, which then becomes

$$I_\nu = \left(\frac{2\pi k^3 T^3}{c^2 h^2}\right) \frac{x^3}{e^x - 1}$$

Differentiating this expression with respect to x with the quantity in the parentheses taken as p, one gets

$$\frac{dI_\nu}{dx} = \frac{3(e^x - 1)x^2 - x^3 e^x}{(e^x - 1)^2}$$

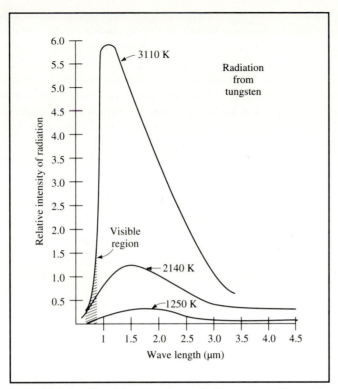

FIGURE 9.1. The blackbody radiation curve. The visible region is the narrow hatched region at the left.

which when set equal to zero yields

$$xe^x = 3e^x - 3$$

from which $x = 2.821$ and

$$T = 2898\frac{\nu_{max}}{c} \quad \mu\text{m-K}$$

This is the *Wien radiation law*, which indicates that ν_{max} is directly proportional to T.

Discrete Sources

A thermal source has a spectrum as shown in Figure 9.1 which is characterized by a continuous distribution of energy. Other sources such as a fluorescent lamp or a mercury street lamp are characterized by much more intense radiation in discrete spectral regions, as shown in Figure 9.2. This figure shows a background continuum on which the intense spectral lines are superimposed; in some cases the background continuum may be absent. To understand more fully the nature of the discrete spectral emission, it is necessary to examine the energy states of atoms.

For an atom or molecule to radiate, it must be in a state where it can give up energy in the form of electromagnetic radiation. Such an *excited state* usually involves the electronic structure of the atom or molecule. Some molecules have energetic vibrational or rotational transitions in the near-infrared region of the spectrum or even in the visible, but typically an electronic transition is required to generate light in the visible region.

Bohr was the first to provide an acceptable theory for the spectrum of the hydrogen atom, where the spectrum is characterized by distinct series of lines beginning in the ultraviolet region and spanning the visible spectrum. Table 9.1 lists the lines of the first three series known as the Lyman, Balmer, and Paschen series, only the Balmer being in the visible spectrum. Bohr postulated that the electron in the hydrogen atom could only occupy certain discrete energy levels or states and that the absorption or emission of radiation by a hydrogen atom was accompanied by a change of state of the atom. Thus a series of lines in the hydrogen spectrum consists of transitions from higher levels to some given lower level. The states or energy levels are characterized by a quantum number n which specifies its level, the lowest level, $n = 1$, being termed the *ground state*. The

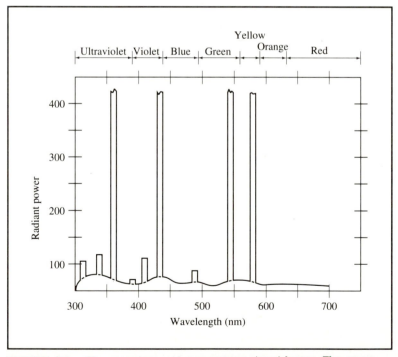

FIGURE 9.2.   The spectrum of a mercury street lamp. The spectrum has both a discrete set of lines and a background continuum.

TABLE 9.1   The Lines of the Hydrogen Spectrum
in the First Three Series[a]

Lyman	Balmer	Paschen
1215.66	6562.79	18,751.1
1025.83	4861.33	12,818.1
972.54	4340.47	10,938.0
949.76	4104.74	10,499.8
937.82	3970.07	9,546.2
	3889.05	
	3835.39	
	3797.90	

[a]The wavelengths are given in angstrom units, 1 Å$=10^{-10}$ m, since that is
the traditional way in which spectral lines are presented. 1 Å = 0.1 nm.

ground state is the state in which the atom is normally found, some excitation energy
being required to promote the atom to its higher state. A diagram of the hydrogen transitions is given in Figure 9.3.

At the simplest level, then, the discrete spectrum of a light source can be understood
in terms of the atomic structure. More complex atoms, such as neon, have a considerably
richer spectrum than hydrogen, but nonetheless the spectrum is discrete. What is of
significance, however, is that the spectrum is not richer in the number of lines than it is.

Sodium has 11 electrons, 10 more than simple hydrogen, yet its spectrum is far
from continuous. To examine this it is necessary to look to the more detailed theory of an
atom's electron energy states which has developed from quantum theory.

The spectrum of more complex atoms such as sodium consists of discrete lines, but
these lines often have additional structure and may appear as pairs or triplets or even more
complex multiplets of lines. The characteristic yellow light of sodium is actually due to a
yellow doublet at 5895.93 and 5889.86 Å. One needs a relatively high dispersion to
resolve these lines. Sodium is similar to hydrogen as a Group I element on the periodic
table, and its spectrum again consists of series of lines; however, in sodium these lines
have additional structure.

According to quantum theory each electron within an atom is characterized by four
quantum numbers. The first, *n*, the *principal quantum number*, is the same as that put
forward by Bohr and is the most important of the quantum numbers in establishing the
energy of the electron's state. The second, the *angular momentum quantum number*, *l*, is
related to the orbital angular momentum of the electron and provides the energy differences which give rise to the structure of the lines such as that of sodium. While *n* takes on
integral values 1, 2, 3, 4, . . . , *l* takes on values 0, 1, . . . , *n* − 1. This means that for
n = 1 there is a single *l* value of 0, but an electron with *n* = 3 has possible *l* values of
0, 1, or 2. Traditionally, the *l* values are denoted by letters rather than numbers, these
being *s*, *p*, *d*, *f*, *g*, so that *l* = 0 is called an *s* state while *l* = 3 is called an *f* state.

There is an additional structure to the angular momentum states which appears when
the radiating atom is in a magnetic field. This arises from the third quantum number, *m*,

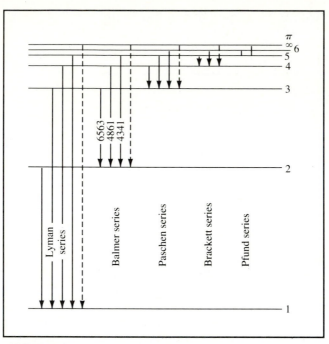

FIGURE 9.3.   A diagram of the radiative transitions of the hydrogen atom.

the *magnetic quantum number*, which takes on values $-l,\ -l + 1,\ \ldots,\ 0,\ 1,\ \ldots,$ $l - 1,\ l$. For an *f*-state electron where $l = 3$, the possible values of m are $-3,\ -2,\ -1,\ 0,$ 1, 2, and 3. The final quantum number, σ, relates to the spin of the electron and has but two possible values, $+1/2$ or $-1/2$. The electron spin couples with the orbital angular momentum to establish the total angular momentum of the electron, which is itself quantized and given the designation j.

For an *s* state $(l = 0)$, j is simply $+1/2$ or $-1/2$ and for a *d* state $(l = 2)$, j is 5/2 or 3/2 $(2 + 1/2$ or $2 - 1/2)$. When a number of electrons are involved in setting the energy state of the atom as is the case when more than one electron is found outside closed electron shells in the ground state, one uses $J = \Sigma\ l + N/2$, where N is the number of electrons outside closed shells in the atom.

The electronic structure of an atom is specified by listing the four individual quantum numbers for each electron in the atom beginning with the lowest energy state, $n = 1$, and successively adding electrons until the number reaches the atomic number of the atom. This is not a random process but a precisely defined one made precise by the Pauli exclusion principle, which states that no two electrons in a single atom can have the same set of quantum numbers. Following this principle the electronic structure of sodium is given in Table 9.2. The procedure is a simple hierarchical one, each successive electron occupying a higher energy state.

TABLE 9.2   The Electron State in Sodium^a (Na)

n	l	m	σ
1	0	0	$+1/2$
1	0	0	$-1/2$
2	0	0	$+1/2$
2	0	0	$-1/2$
2	1	1	$+1/2$
2	1	0	$+1/2$
2	1	-1	$+1/2$
2	1	1	$-1/2$
2	1	0	$-1/2$
2	1	-1	$-1/2$
3	0	0	$+1/2$

^{a}It should be noted that the electron spins are added with parallel spins as far as possible.

In the case of sodium the final electron added goes into the $n = 3$ state as the only electron with $n = 3$. As a result sodium has a spectrum somewhat akin to that of hydrogen in that it has but a single electron outside completed shells, although it is richer in lines than hydrogen. The single $n = 3$ electron is why sodium appears below hydrogen in the periodic table of the elements.

Given all the possible states immediately above the ground state of sodium, it may seem surprising that the spectrum is not very much richer in lines. The reason the spectrum is not richer is the existence of *selection rules* that apply when an atom is radiating. As was shown for hydrogen, a transition between any two unoccupied levels of different n is permitted and gives rise to a spectral line. The same is true for more complex atoms with the additional constraint that J can only change by ± 1 in a radiative transition; no other values are possible. Thus if J is 5/2 for an electron in an excited state, the state may radiate and the electron fall to a lower state but only to a state where $J = 7/2$, 5/2, or 3/2. All other transitions would be forbidden. This is illustrated in Figure 9.4. Forbidden transitions do occur but at a very much diminished rate compared with allowed transitions; the time scale may vary by a factor of as much as 10^5. The doublet structure of the sodium spectrum arises from transitions shown in Figure 9.4. The states are labeled with their principal quantum number, 3 in this case, their l value as s, p, d, and so on with a superscript prefix specifying the sum of the spins as $(2S + 1)$ or 2 in the case of a one-electron atom like sodium; these expressions are called *spectral terms* and are indicative of the energy of the state. Additionally, a subscript suffix is given to specify J. No transition between the $3^2D_{5/2}$ and the $3^2S_{1/2}$ ground state is allowed. The superscript prefix of the state is indicative of the multiplet structure of the spectrum, so that one finds sodium to have a doublet spectrum as indicated earlier. Calcium with two electrons outside closed shells has $S = 1/2 + 1/2 = 1$ and is a triplet, $2S + 1 = 3$, state. The spectrum of calcium is characterized by closely spaced triads of lines.

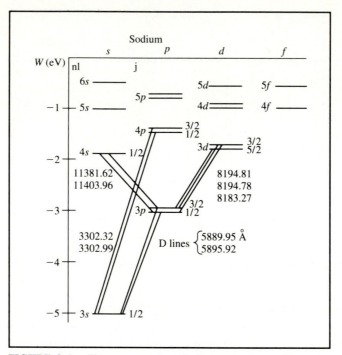

FIGURE 9.4. The energy-level diagram for sodium with some of the permitted transitions shown with their wavelengths.

EXAMPLE 9.2   Give the complete electronic structure of calcium, element 12 on the periodic table of the elements, and give its ground-state spectral term.

Solution   The electronic structure is

n	l	m	s
1	0	0	$+1/2$
1	0	0	$-1/2$
2	0	0	$+1/2$
2	0	0	$-1/2$
2	1	1	$+1/2$
2	1	0	$+1/2$
2	1	-1	$+1/2$
2	1	1	$-1/2$
2	1	0	$-1/2$
2	1	-1	$-1/2$
3	0	0	$+1/2$
3	0	0	$-1/2$

The filling of shells proceeds as indicated, with electrons being placed in levels with parallel spins until no further parallel spins can be added. The last two ($n = 3$) electrons are outside complete shells, and these form the ground-state spectral term. $S = 1/2 - 1/2 = 0$ and $J = L + S = 0$, so that the ground state is $^1S_0$. Excited states can have parallel spins giving $S = 1$, and these would be triplet states. The paradox of many electrons and a relatively sparse spectrum is then resolved by the selection rules. These rules really are conservation laws applied to microscopic systems and as such are to be expected.

Emission and Absorption—The Einstein Coefficients

The wavelength of the radiation emitted by an atom in a transition from state E_1 to state E_0 (or absorbed in going from E_0 to E_1) is governed by Planck's law.

$$E_1 - E_0 = \frac{hc}{\lambda} \tag{9.4}$$

While this law clarifies the relationship between energy states and the emitted wavelength, it does not give any hint about what processes are involved in the transition or the time scale on which such processes occur.

The well-known Boltzmann factor gives the equilibrium distribution of atoms between an excited state E_1 with N_1 atoms and the ground state E_0 with N_0 atoms,

$$\frac{N_1}{N_0} = e^{(E_1 - E_0)/kT} \tag{9.5}$$

where k is Boltzmann's constant and T is the absolute temperature.

EXAMPLE 9.3

What fraction of a population of gas atoms is found in an excited state 1 eV (1 eV = 1 electron-volt = 1.6×10^{-19} J) above the ground state at room temperature, $T = 300$ K:

Solution

$$\frac{N_1}{N_0} = \exp\left[\frac{-1.6 \times 10^{-19} \text{ J}}{1.38 \times 10^{-23} \text{ J/K} \times 300 \text{ K}}\right] \tag{9.6}$$

which gives $e^{-38.6} = 1.64 \times 10^{-17}$. If there is 1 kg-mole of gas atoms present, that is, 6.023×10^{26} atoms, there will be $\sim 10^{10}$ atoms in the excited state.

This example shows that even though there is only a very small fraction of the atoms in the excited state, the population is not insignificant. How do these atoms return to the ground state? Two processes are involved. First, and usually predominantly, the transitions are mechanical. The atoms collide with each other and the walls of the container and give up their energy to other atoms or to the walls. In the case of other atoms, these atoms then enter the excited state since atom–atom collisions are nearly perfectly elastic. The second process is spontaneous radiation. The atom gives up its excess energy to the electromagnetic field and produces radiation that obeys Planck's law (9.4).

The latter process is termed *spontaneous emission* and is characterized by a coefficient A. The amount of spontaneous emission in an excited state E_1 with N_1 atoms is

$$N_i A \text{ photons/s} \tag{9.7}$$

These photons must be reabsorbed or gas would cool spontaneously. The reabsorption process is characterized by a coefficient B_{01} for atoms going from the ground (0) state to the excited (1) state. The rate of absorption of this radiation is given by

$$N_0 B_{01} U_\lambda(\lambda) \text{ photons/s} \tag{9.8}$$

where $U_\lambda(\lambda)$ is the density of electromagnetic energy corresponding to the $0 \rightarrow 1$ transition. Thus there are two excitation processes: mechanical and radiative.

In considering the problems associated with radiating objects, Einstein reasoned that if there was an excitation transition due to the radiation field in the gas, there must also be a deexciting transition due to the field, and he postulated

$$N_1 B_{10} U_\lambda(\lambda) \text{ photons/s} \tag{9.9}$$

for this downward transition, where B_{10} is the coefficient for the excited (1) state to the ground (0) state transition. If one notes that the mechanical transitions are essentially in equilibrium, then the radiative transitions must be also, and equating these one gets

$$N_1 [B_{10} U_\lambda(\lambda) + A] = N_0 [B_{01} U_\lambda(\lambda)] \tag{9.10}$$

This equation can be solved for N_1/N_0, which can then be replaced by the Boltzmann factor from (9.5):

$$\frac{N_1}{N_0} = e^{E_1 - E_0/kT} = \frac{B_{01} U_\lambda(\lambda)}{B_{10} U_\lambda(\lambda) + A} \tag{9.11}$$

This expression can be rearranged to give $U_\lambda(\lambda)$, where $E = E_1 - E_0 = h\nu$:

$$U_\lambda(\lambda) = \frac{A}{B_{01} e^{h\nu/kT} - B_{10}} \tag{9.12}$$

Comparison of this with Planck's law (9.3) gives

$$B = B_{01} = B_{10} \quad \text{also} \quad \frac{A}{B} = \frac{8\pi h\nu^3}{c^3} \tag{9.13}$$

One important factor in the radiation-induced emission called *stimulated emission* is the fact that the radiation which is emitted by an atom radiating under this process is in phase with the field inducing the emission. While the spontaneous emission characterized by the Einstein *A* coefficient can occur at any time and in any direction, the stimulated emission has the wave vector and phase of the stimulating radiation.

The development used by Einstein was based on an equilibrium assumption. In an equilibrium state there can be no more atoms in the excited state than given by the Boltzmann faction $e^{-h\nu/kT}$. As a result there is always a greater potential for the absorption of any radiation than for emission, since the Einstein *B* coefficients are the same for absorption and emission. Only in the past three decades have ways been found to produce an excess nonequilibrium population in the excited state and thus produce a gain in light energy. In the presence of a very intense radiation field at the transition frequency ν, one no longer has an equilibrium condition, and the population varies from that predicted by the Boltzmann factor. The population of atoms becomes *saturated*; there are as many absorption transitions as emission transitions and the state populations are equal. To achieve an inverted population with more atoms in the excited state than in the lower energy state, one must look to processes other than a simple intense radiation field.

Lasers

The previous paragraph indicated that even in a very intense radiation field the number of atoms in the excited state does not exceed the number in the ground state. What would happen if the upper state were more highly populated than the ground state? In this case there would be a *gain* in the radiation field. More light would come from the atomic population than was incident upon it. Devices which have such a population inversion and a radiation gain are called *lasers*, an acronym for light amplification through stimulated emission of radiation. In this section, several of the most widely used lasers will be described. Today there are literally hundreds of laser sources, each unique in that each one has a special procedure for achieving population inversion.

Most lasers can be broadly classified as three-level systems or four-level systems. While such a classification is a rather simplified approach, it will serve to illustrate the laser process; there can be literally hundreds of energy levels involved in some laser systems. Figure 9.5 illustrates these processes. In both cases the lowest state is the ground state or another similarly highly populated state. The production of an inverted population is called *pumping*. The pumping action in both cases is to a higher state not involved in the laser transition. The state to which the pumping occurs decays to an intermediate state labeled 2 in the figure. This decay must be very rapid to prevent a buildup in state 1. State 2, which is populated by the decay from state 1, is characterized by a small Einstein *A* coefficient relative to the decay $1 \rightarrow 2$, which has a much larger coefficient. As a result an excess population is developed in state 2. The laser transition is then from state 2 back to the ground state 0 in the three-level system in Figure 9.5a and to a lower state 3 in the four-level system (b). State 3 must empty quickly in the four-level system to prevent the quenching of the laser process.

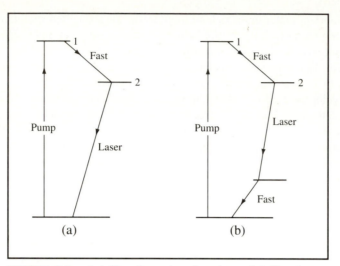

FIGURE 9.5. (a) The three-level system. (b) The four-level system.

The Helium–Neon Laser

The helium–neon or He–Ne (pronounced "hee-knee") laser was one of the earliest lasers. Even considering its pioneer position, it remains the most widely used and widely available laser today. It was developed by Ali Javan in 1960 and had an output in the infrared at 1.15 μm. Today most He–Ne lasers operate at 632.8 nm in the red portion of the visible spectrum. Other laser transitions are possible with this laser as well, for example, in the green at 543.36 nm, and there are more than 100 possible transitions in the neon which can exhibit laser action.

The pumping process for the He–Ne is a result of the accidental coincidence between the energy levels of an excited state of helium and that of neon. Actually a number of such coincidences occur, two of which are illustrated in Figure 9.6.

The He–Ne laser is representative of the gas-discharge lasers which use gases and which achieve their population inversion through the action of an electrical current flowing through a tube of gas. Gas-discharge lasers have a highly homogeneous laser medium because of the randomizing action of the gas-discharge process. They are also very efficient. The output beam is a continuous wave (cw) output, in contrast to the pulsed outputs of many other laser systems. As can be seen from the figure, this laser is a four-level system.

The typical He–Ne laser as shown in Figure 9.7 consists of a tube from 10 cm to 1 m in length and approximately 5 mm in bore diameter. It is filled with a mixture of about 10 parts helium to 1 part neon at a total pressure of about 0.2 mm Hg. Scaling rules are used when one changes the bore of the tube, with the filling pressure being inversely

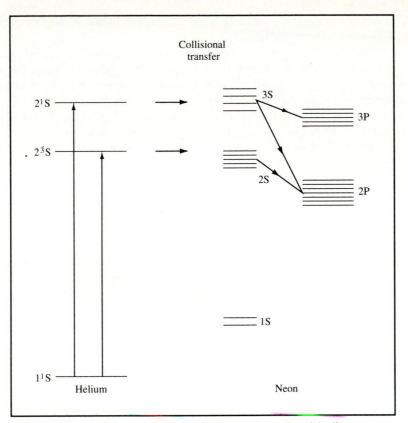

FIGURE 9.6. A schematic of the transitions involved in the He–Ne laser.

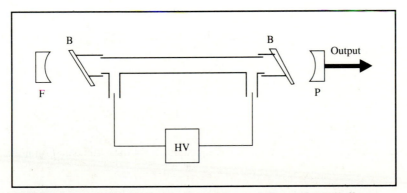

FIGURE 9.7. A He–Ne laser. F and P are fully and partially reflecting mirrors, respectively; B indicates the Brewster windows; and the high-voltage supply is shown as HV.

proportional to the bore. The tube gases are excited by an electrical current of about 50 mA flowing through the tube.

The electrons accelerated through the gas as a result of the initial breakdown of the gas in the electric field collide with the helium atoms and excite them. The voltage across the tube depends on length but is usually of the order of 5 kV, which is more than sufficient to initiate the breakdown of the gases. A significant fraction of the helium will be excited to the 2^1S state or a higher state and decay to the 2^1S state. This state is metastable with respect to the ground state of helium so that the decay back to the ground state will be comparatively slow. As a result, the atoms in this state decay primarily through mechanical exchanges in collisions with other atoms or the walls. The 3S states of the neon atoms have an energy closely coincident with this metastable state of helium. The 3S neon states are then populated directly by the gas discharge, but more importantly, by collision with metastable helium. The result is an inverted population in which the populations of the 2S and 3S states exceed those of the lower-energy 2P and 3P states; a population inversion exists. The candidates for laser transitions are those spectral transitions with a relatively small A coefficient and these spectral transitions are known as weak lines. The transition must end on a state with a large A coefficient, a strong line, so that the lower state of the transitions empties quickly so that the population inversion can be maintained. In this gas mixture these conditions are met.

The ends of the laser tube are Brewster windows which by their orientation preferentially select one of the polarization directions of the laser output for transmission. The entire tube is placed between two mirrors, one with nearly total (99.9%) reflectivity and the other with a reflectivity of about 99.0%. These mirrors form a resonant cavity so that the field within the discharge tube is coherent, with a standing wave being set up within the tube. The condition for this is that the optical length of the tube is an integral number of half-wavelengths

$$r\frac{\lambda}{2} = nl \tag{9.14}$$

where r is a positive integer and n the refractive index of the medium. In terms of the frequency ν of the radiation

$$\nu = \frac{c}{\lambda} = \frac{rc}{2nl} \tag{9.15a}$$

and for a fixed l, the separation of frequencies for which this equation holds can be found from

$$\frac{d\nu}{dr} = \frac{c}{2nl} \tag{9.15b}$$

EXAMPLE   What is the separation of frequencies in a 50-cm He–Ne laser? The index of the gases can
9.4   be taken to be 1.

Solution   Using equation (9.15b)

$$\frac{dv}{dr} = \frac{3 \times 10^{10} \text{ cm/s}}{2 \times 50 \text{ cm}} = 300 \quad \text{MHz}$$

The line width of the laser transition is about 1000 MHz, so that there will be more than one resonant condition in the laser cavity, as one can see from the preceding example. These separate coexisting resonance conditions are termed *laser modes*, and they correspond to cavity resonance modes in electromagnetic theory. The solutions of the cavity fields are somewhat complicated here since the mirrors are spherical rather than plane. Spherical mirrors are used since they are much easier to align than the plane mirrors which were originally used. The modes can be observed in the output beam passing through the partially transmissive mirror. The simplest mode gives a uniform beam across the output aperture, while other modes give amplitude minima across the beam as illustrated in Figure 9.8.

With the large number of processes of excitation and deexcitation taking place simultaneously, calculation of the gain of the system is at best complex. Even restricting the processes under consideration to the formation of the metastable state of helium and to the population of the three active states of the neon, one has a complex set of four coupled differential equations to solve. This becomes a mathematical horror! One can, however, easily establish a threshold condition for laser action. One begins with the Beer–Lambert law

$$I = I_0 e^{-\alpha l} \tag{9.16}$$

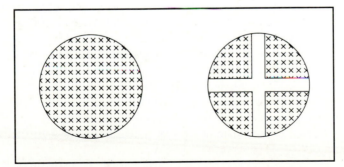

FIGURE 9.8. Two cavity modes for a He–Ne laser. The fundamental is shown on the left while a second mode appears on the right.

where I is the intensity of light with initial intensity I_0 after this light has traveled a length l of a medium characterized by the absorption coefficient α. The loss at the mirrors can be described by a parameter given by

$$\gamma = -\log_e(R_tR_p) \tag{9.17}$$

where R_p and R_t are the reflectivities of the partially and totally reflecting mirrors. After one round trip through the laser

$$I = I_0 e^{-2\alpha l - \gamma} \tag{9.18}$$

and for gain to take place the threshold condition is

$$\alpha = -\frac{\gamma}{2l} \tag{9.19}$$

Since γ is inherently negative, α must be sufficiently large to overcome the mirror losses. Full treatment of the rate equations results in values of α such that the output power of the He–Ne laser is restricted to the order of milliwatts.

　　Other noble gases, mainly krypton and argon, have produced active laser transitions in their ionized states. These require very large tube currents, and while producing a much stronger output of about 5 W cw, the conversion efficiency is only of the order of 0.03%, so that a power supply of about 12 kW is required for them. Their principal use to date has been to serve as pumps for dye lasers.

The CO_2 Laser

The carbon dioxide laser is an important industrial device in that it has a very high efficiency and can achieve hundreds of kilowatts cw or a very much higher pulsed output. The CO_2 laser is unlike the He–Ne laser in that CO_2 is a molecular gas and it has, in addition to electronic states, a large set of lower-energy rotational and vibrational energy states. Just like the electronic states, these are quantized. The energy levels of the vibrational and rotational states are such that the transitions between them are in the infrared. The typical CO_2 laser output is at a wavelength of 10.4 μm.

　　The CO_2 molecule is a symmetrical, linear molecule with the carbon atom set between two oxygen atoms. Figure 9.9a shows the geometry of this molecule. In addition to rotational states, this molecule has the vibrational states illustrated in Figure 9.9b–d. Again with this laser it is the accidental coincidence of energy states which provides the pumping. The first vibrational level of the nitrogen molecule is nearly coincident with the asymmetric stretching excited state of CO_2 as shown in Figure 9.10.[1] It is notable that

---

[1] In the infrared spectrum the energies are commonly expressed in units of cm^{-1}. These can be converted to more conventional units by multiplying them by hc, where h is Planck's constant and c is the velocity of light. The vibrational and rotational energies are of the order of 1/100 those of electronic energy states.

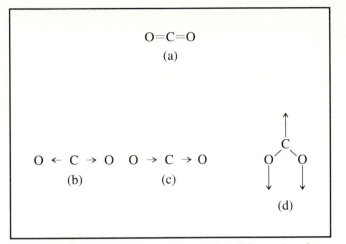

FIGURE 9.9. (a) The geometry of the CO_2 molecule. (b) The symmetric stretching vibration. (c) The asymmetric stretching vibration. (d) The bending vibration.

radiative transitions between vibrational states of nitrogen are forbidden since the molecule has no dipole moment, so that once the vibrational state of nitrogen is excited, it can transfer its energy principally by the collisional process. The CO_2 laser is one example of a four-level laser system.

The apparatus for the CO_2 laser is similar to that in Figure 9.7. The high-voltage source causes the gas mixture in the tube to break down, and the electrons formed accelerate along the tube and collisionally excite the gases present. The CO_2 laser is very efficient since the states excited by these collisions are low energy and, therefore, easily excited. A mixture of CO_2, N_2, and He is used in the laser; the presence of the He is required to help depopulate the 667-cm^{-1} state to prevent quenching of the process. Collisional excitation to all the states shown in Figure 9.10 takes place.

The CO_2 laser is usually a flowing gas system, since the CO_2 can decompose in the discharge into CO and ultimately into C and O. This can result in a buildup on the tube and windows, so the gas mixture is made to flow through the tube to sweep out these decay products. In sealed-tube CO_2 lasers, water is added to the gas mixture to help prevent formation of CO.

The Ruby and Nd:YAG Lasers

The first laser to exhibit laser action in the visible region of the spectrum was the ruby laser, which uses a rod of ruby crystal as the laser medium and which has its output in the red at 694.3 nm. The solid-state active medium is quite different from the gaseous media in the He–Ne and CO_2 lasers.

Ruby is a sapphire crystal, a crystalline structure of alumina (Al_2O_3), to which a small doping of 0.05–1% of Cr^{3+} as Cr_2O_3 has been added. The ruby crystal has two

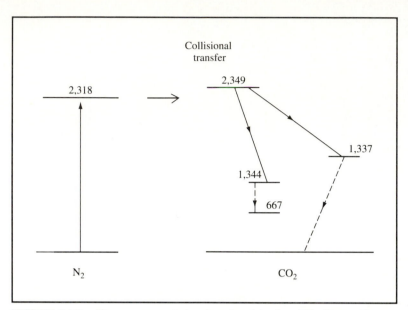

FIGURE 9.10. The energy states involved in the CO_2 laser. The solid downward lines are the laser transitions; the dashed lines are not laser transitions.

absorption bands, one in the blue and the other in the green, each approximately 100 nm wide. These bands account for the red color of the ruby since the reflected light from the ruby is principally in the red. Figure 9.11 shows some of the features of the spectrum of the ruby crystal. In addition to the absorption bands, there is a pair of very closely lying states separated by 29 cm^{-1} which form the upper state for the laser transition which terminates on the ground state.

As a solid, the lattice acts like a large molecule, so that with the Pauli principle, no two electrons may have the same set of quantum numbers, and this ultimately results in the excited states of the aluminum ions being separated by a small energy. The bands are made up of a very large number of discrete states so close in energy that they may be thought of as continuous. The electrons excited into the green band have a lifetime of the order of 10^{-7} s, and these decay, transferring their energy to the chromium ion state labeled R_1 in the figure. This state is metastable with a lifetime of 4.3×10^{-3} s. The resulting spectrum is precisely that required for a three-level laser system.

The ruby laser is optically pumped using a xenon flash lamp, and the ruby rod, usually about 5 cm long and 1 cm in diameter, is set at one focus of an elliptical, cylindrical mirror in which the flash tube is set at the other focus as in Figure 9.12. When the flashlamp is fired, the light from the flash tube which strikes the surface of the mirror will be directed onto the ruby rod, since the foci of an elliptical mirror are conjugate with each other. The high energy necessary to fire the flash tube and the high temperature required to pump the absorption bands necessitates that the ruby laser be operated in a pulsed mode rather than continuously. The efficiency of this laser is about 0.7%.

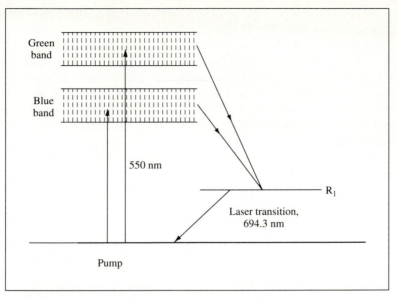

FIGURE 9.11. The salient features of the ruby crystal spectrum for its action as a laser medium.

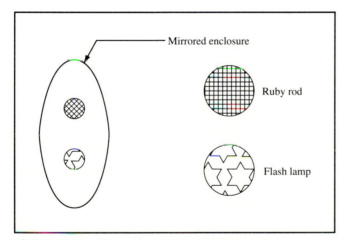

FIGURE 9.12. The ruby laser pumping system.

The ends of the ruby crystal are cut so that one is at the Brewster angle and a partially transmitting mirror is set beyond this end of the rod which serves as the output end of the laser. The other end is cut as a corner cube so that the end is totally reflecting. When the flash fires, electrons are excited to the green band, and these rapidly fill the R_1 level, creating an inverted population, and laser action commences. The output pulse duration is about 10^{-6} s with an energy output of 10^{-2} J. While this may seem small, it should be noted that the output is 10 kW!

In practice it is possible to increase the output of a pulsed laser through a process known as *Q-switching*. The Q refers to the quality factor of the resonant cavity.[2] Laser action will begin in the laser as soon as population inversion takes place and will continue at that level until the inversion is depleted. If the output mirror is cut off from the system while pumping is taking place, the cavity is "spoiled," and a greater inversion can be gotten since the rate of depletion of the laser state is less than it would be as part of a resonant cavity. Once the spoiler is removed, the laser can function with a much greater gain than would otherwise be the case. One can generate pulses with a 10 ns width and an energy of about 0.1 J.

Q-switching is done in several ways. One can rotate the output mirror rapidly and synchronize its position so that a cavity is formed only when the population has gotten its maximum inversion. One may place a dye in a cell in the light path between the Brewster window and the output mirror. By properly selecting the dye and setting its concentration, the dye can initially absorb the 694.3-nm light spoiling the cavity, but it quickly becomes saturated and thus transmitting when the population has achieved its maximum inversion. Finally there are electro-optical shutters which can be opened rapidly in proper synchronization with the flash pulse.

Another crystal laser which is important today is the neodymium YAG laser. Yttrium aluminum garnet (YAG) is a crystalline host material $Y_3Al_5O_{12}$ that can be doped with the Nd^{3+} rare earth ion at a level of 1–2%. Nd^{3+} is also used as a dopant in glass but the Nd:YAG seems to be the dominant laser crystal today. The laser is a classical four-level system as shown in Figure 9.13.

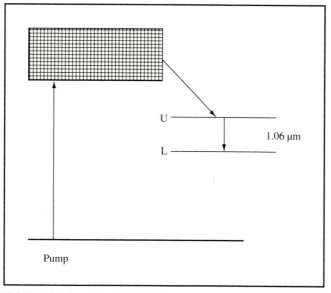

FIGURE 9.13. The Nd:YAG laser.

---

[2]The quality factor is the ratio of energy stored to energy lost per second.

The laser is usually flashlamp pumped and operated at a high pulse rate. The lower energy state in the laser transition decays with a lifetime of about 30 ns but to a radiation which is strongly absorbed by the host lattice. This results in heating of the laser medium, but this heat can be rapidly dissipated so that this laser can even be operated cw if such is required. The output is at 1.06 μm and this has been found to be a useful device for laser surgery.

Laser Diodes

An important class of lasers which finds ever-increasing use in communication systems is the semiconductor laser called a *laser diode*, or an *injection laser*. This is closely related to the *light emitting diode* (LED), which is widely used in various display systems.

Light emitting diodes are formed as a *pn* junction in a semiconducting material. Compound semiconductors such as gallium arsenide (GaAs) usually make better emitters. Table 9.3 lists a number of semiconductor compounds which have been used to make LEDs.

Like other materials, the output of one of these semiconductor materials depends upon it electronic structure. The electronic levels of solids are made up of bands of energy levels as was the case with ruby. The characterization as metal, semiconductor, or insulator is dependent upon the ways the electrons fill the bands. Figure 9.14 illustrates this. The lowest energy band containing unoccupied states is called the *conduction band*, while the uppermost band normally populated with electrons is called the *valence band*. In metals the conduction band and the valence band overlap. For an electron to move through the solid to give conduction, it must be excited to a higher energy state, a relatively easy process in metals, where these states are very close together. In a semiconductor the conduction band lies above the fully occupied valence band, the separation of the bands being called the *bandgap*, E_g. At normal temperatures some electrons will be in the conduction band, as one can predict from the Boltzmann factor (9.5) and a knowledge of the bandgap. The electron population is usually such that the electrons cannot carry large currents as can a metal. Finally, insulators are characterized by a relatively large bandgap and a vanishingly small population of the conduction band with electrons.

TABLE 9.3 Semiconductor Materials for LEDs

Material	Bandgap	Output
ZnS	3.8	UV
ZnO	3.2	UV
CuCl	3.1	UV
ZnSe	2.7	Visible
CdS	2.52	Visible
ZnTe	2.3	Visible
CdSe	1.75	IR-visible
CdTe	1.5	IR
GaAs	1.45	IR

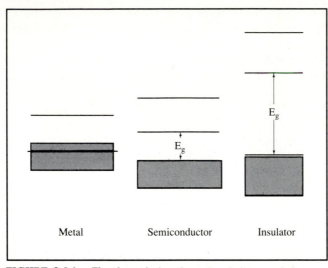

FIGURE 9.14.   The band structure for (a) a metal,
(b) a semiconductor, and (c) an insulator.

The conduction process in semiconductors involves two kinds of charge carriers: the electrons, which are in the conduction band, and the *holes*, the unoccupied levels in the valence band that act as positively charged charge carriers. The effect of this is to alter the Boltzmann factor for semiconductors so that it becomes

$$e^{-E_g/2kT} \tag{9.20}$$

for these materials.

The conducting properties of semiconductors can be altered by a process called *doping*. Pure semiconductors are termed *intrinsic* and are characterized by an equal number of positive (hole) and negative (electron) charge carriers. If a small amount of dopant with a different valence is added, such as adding phosphorus to silicon, the excess electron in the phosphorus becomes a charge carrier which is relatively free to move through the lattice. The material itself is charge neutral but a flowing current will have more electrons than holes and will be called *n-type*. Similarly addition of boron will lead to an excess of holes and a *p-type* material with more positive than negative charge carriers.

Now what happens when a *p*-type and an *n*-type material are joined? Electrons from the *n*-type material will collide with and neutralize holes from the *p*-type material in the region of the junction. As this recombination takes place, the *n*-type side of the junction will develop a positive potential from having lost some negative carriers, while the opposite will be true for the *p*-type side of the junction. The result will be to retard further recombination, since electrons will not easily move away from the positive potential, and on the other side of the junction the holes will remain more closely bound due to the negative potential. The bands in the region of the junction will then be distorted as shown in Figure 9.15.

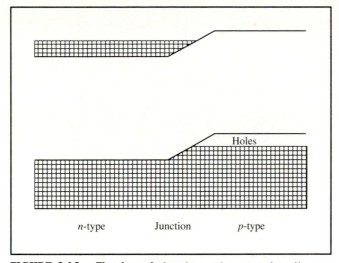

FIGURE 9.15.   The band structure of an *n-p* junction.
Recombination of the electron–hole pairs results in a
potential barrier across the junction. Electrons appear
in the conduction band in the *n*-type regions and
holes in the valence band in the *p*-type region.

Diode action occurs when an alternating voltage is applied across the device as in
Figure 9.16. When the potential on the right-hand *p*-side is positive, holes are forced
across the junction while electrons are similarly driven out of the *n*-region and a current
flows. This is called forward biasing. When the potential at the right-hand side is relatively negative (reverse biasing), holes are drawn away from the junction, as are the
electrons at the *n*-side, and as a result no current flows.

When the device is forward biased, not all the electrons and holes pass completely
through the device; some electrons and holes come together and recombine. This recombination results in a release of energy equal to the bandgap, and the energy appears in the
form of radiation called *recombination radiation*. For many semiconductors, this occurs
in or near the visible. Table 9.3 lists but a few of these.

The recombination radiation in these devices is at a wavelength given by

$$\lambda = \frac{hc}{E_g} \tag{9.21}$$

where E_g is the bandgap, the energy separation of the valence band and the conduction
band. When E_g is given in electron volts, the wavelength in nanometers is given by

$$\lambda = \frac{1240}{E_g} \tag{9.22}$$

so that for ZnSe with a gap of 2.7 eV, the output wavelength is 459 nm in the blue part of
the visible spectrum.

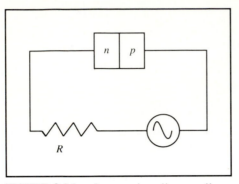

FIGURE 9.16. An n-p junction recti-
fier.

The simple *p-n* junction is not an efficient device for the generation of light and is not used as a light emitting diode. The light generated by the recombination radiation is radiated in all directions and comes out at the end and edges of the device. With such a device, known as a *homojunction*, only a small fraction of the output light can be used.

A more efficient device can be made using a *heterojunction* of two dissimilar semi-conductors with different bandgaps. Both electrons and holes must cross a boundary to interact and recombine only in a small region, the *confinement region*. The dissimilar materials also form a light pipe in the central confinement region, directing the output in a useful direction. The band structure of a heterojunction is shown in Figure 9.17.

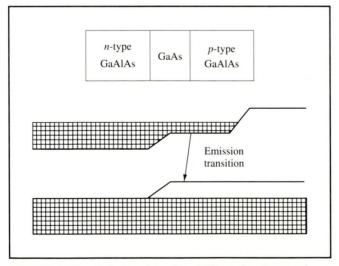

FIGURE 9.17. A heterojunction LED under forward
bias.

Light emitting diodes are made in two forms, edge emitters and surface emitters. These are illustrated in Figure 9.18. The surface emitter called an etched-well or Burrus diode is less efficient than the edge emitter for fiber optic coupling since its output cone is too wide for full utilization. Both types of devices do find extensive use, however.

The output light energy is linear in the current in an LED. This can be seen in Figure 9.19. This linear relationship depends on the rate of recombination, which is a constant η times the number of charge carriers. One has

$$\text{power} = \eta N E_g \tag{9.23}$$

If E_g is in electron volts and one notes that $N = i/e$, then

$$\text{power} = \eta E_g i \tag{9.24}$$

and the linear relationship is clear.

Laser diodes are nearly identical to LEDs except that laser diodes are only edge-emitters. The population of electron–hole pairs available for emission is controlled by the current. The output power–current relationship is shown in Figure 9.20. Once the current has reached a critical level, laser action begins. The laser diode is typically only a few tenths of a millimeter long, and the surfaces of the diode have a high enough reflectivity so that no external mirrors are required. This can be seen if one uses (7.20) and notes that for the typical semiconductor $n = 3.5$, the reflectivity is 31.36% of the light incident on the air–semiconductor interface. Good planar surfaces are gotten by cleaving the crystals along crystal planes to ensure good surface reflectivity.

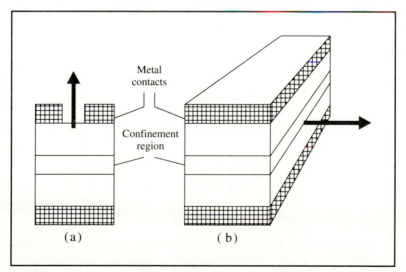

FIGURE 9.18.   (a) A surface-emitting LED. (b) An edge-emitting LED.

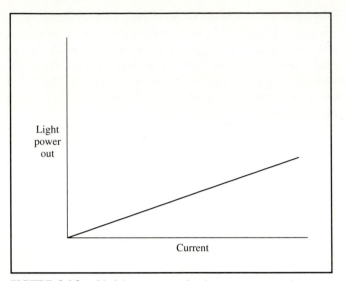

FIGURE 9.19.   Light power output versus current for an LED.

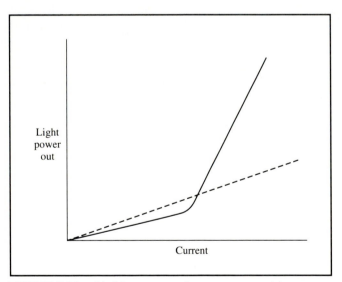

FIGURE 9.20.   Light power out versus current for a laser diode. The dashed curve represents the LED output.

Laser diodes are usually operated in a pulsed mode. One reason for this is to control the heating of the diode. Too much current can destroy a semiconductor device. As it is, the threshold current for laser action depends on the temperature of the diode. Laser diodes are often operated chilled to liquid nitrogen temperatures (77 K), which results in nearly an order of magnitude increase in radiation efficiency.

Laser diodes are extremely important in fiber optic communication systems. Such systems will be discussed in the next chapter.

PROBLEMS

9.1. What are the units of radiation intensity in equation (9.1)? How much will the output intensity of a blackbody source increase if its temperature initially at 300 K (room temperature) is increased by 10 K?

9.2. At what temperature will the output maximum of a radiating body be at 550 nm? at 600 nm? How does this change with ϵ?

9.3. A toaster wire made of nichrome with $\epsilon = 0.91$ is 1.0 m long, has a diameter of 1.5 mm, and is maintained at 750°C. How much power is radiated from it?

9.4. Show that argon, element 18, is inert as a result of having only completed electron shells.

9.5. Give the quantum numbers of each electron in aluminum, element 13.

9.6. What is the energy difference between the states which give rise to the sodium D-line at 589.6 nm?

9.7. What are the upper and lower limits of the transition energy within atoms and molecules which can generate visible light? Take the limits of the visible spectrum as 400 nm and 700 nm.

9.8. Using Planck's law with λ as the variable rather than frequency and noting that $I_\lambda d\lambda = I_\nu d\nu$, derive equation (9.1).

9.9. If B_{10} is 10^{18} m³/W, what is A_{10} for a wavelength of 400 nm?

9.10. What is the separation of resonant frequencies in a 10-cm He–Ne laser? a 100-cm laser? Why might one be preferable to the other?

9.11. A He–Ne laser has a tube 50 cm long. One window has reflectivity of 0.9996 and the other 0.989. What value of α will result in gain?

9.12. Based on the bandgap energy in Table 9.3, estimate the output wavelength for a ZnO, a CdS, and a CdTe LED. How does this answer relate to that of Problem 9.7?

CHAPTER 10

Fiber Optics and Fiber Optic Systems

Along with lasers, one of the most important advances in optics in recent years has been the introduction of optical fiber communication systems. Over the next few decades copper line communication links will be largely replaced by optical fibers. These fibers provide the potential for much greater bandwidth in the communication links and will likely change the nature of many of the communication practices currently in use.

The fact that light could be guided by use of total internal reflection in various media was demonstrated in a stream of water by Tyndall in 1870. The practical application of this effect has awaited other technology, which in the past 30 years has begun to come into place. Fiber communication links are now coming on line at a rapid rate. Recently a transoceanic fiber link between the United States and Great Britain has been established.

Fiber optical systems are similar to other communication systems. At the transmitting end an electrical signal is converted into a modulated light signal which is then transmitted along a fiber optic line. At the receiving end the optical signal is demodulated and reconverted into an electrical signal. One advantage of such a system lies in the fact that the frequency of light is about 10^4 times that of the present microwave systems; thus an optical system can in principle carry 10^4 times the information as a microwave system. This is but one of the significant features of fiber systems. Other advantages (as well as problems) exist and will be presented throughout this chapter.

The Optical Fiber

One source of impetus in the development of fiber communication systems has been in the fibers themselves. Table 10.1 shows how changes have occurred in fiber attenuation as a result of improvements in both material and manufacturing.

TABLE 10.1   Optical Fiber Attenuation[a]

Year	Attenuation (dB/km)
1960	1000
1970	20
1973	4
1974	2
1979	0.2

[a]Limiting value ~0.1 dB/km.

There are two important kinds of optical fiber: step-index fiber and graded-index fiber. The step-index fiber has an index profile as shown in Figure 10.1a, whereas the profile of the graded-index fiber is shown in Figure 10.1b. The step-index fiber can be either glass with glass or plastic cladding or plastic with plastic cladding. An outer jacketing layer is often added to protect the fiber mechanically. As can be seen in Figure 10.1a, the center strand of the fiber, the *core*, has a higher index than the cladding to make use of total internal reflection at the interface, thereby constraining the optical signal to the core. A single, unclad fiber would be susceptible to light leakage if the surface were scratched or handled, leaving oils on the surface.

The dimensions of the core and cladding are important in fixing both optical and physical characteristics of the fiber. Small-diameter fibers have a larger bandwidth but are difficult to splice and handle. Cores range from 5 μm to 0.5 mm, while the cladding diameters are generally between 0.1 and 0.7 mm. An additional 0.1 mm is added to the overall diameter of the system by the protective sheathing.

Figure 10.2 shows the coupling of a light source to an optical fiber. Only light within a narrow cone will propagate in the fiber. This cone is known as the *cone of acceptance*, and light falling on the end of the fiber outside this cone of acceptance will be lost. The cone of acceptance is easily defined by taking a core index of n_c and a cladding index n_{cl} with $n_{cl} < n_c$. Figure 10.3 describes the geometry. Let θ_a be the apex semiangle of the cone. The angle θ_t is that angle at which one has the onset of total internal reflection on the interface between the core and cladding layer. This is given by

$$\theta_t = \sin^{-1}\left(\frac{n_{cl}}{n_c}\right) \tag{10.1}$$

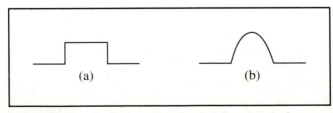

FIGURE 10.1.   Index profiles for (a) the step-index fiber and for (b) the graded-index fiber.

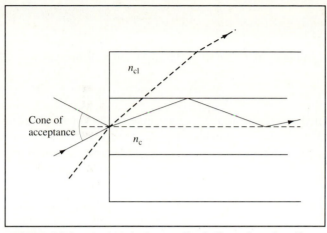

FIGURE 10.2. A step-index optical fiber showing the cone of acceptance.

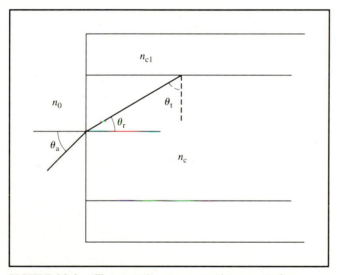

FIGURE 10.3. The maximum acceptance angle.

using Snell's law, equation (2.1). The angle of refraction θ_r for an input at θ_a is $90° - \theta_t$ and

$$\sin \theta_r = \cos \theta_t = \cos\left[\sin^{-1}\left(\frac{n_{cl}}{n_c}\right)\right] \tag{10.2}$$

so that

$$n_0 \sin \theta_a = n_c \sin \theta_r = n_c \cos\left[\sin^{-1}\left(\frac{n_{cl}}{n_c}\right)\right]$$

or

$$n_0 \sin \theta_a = \sqrt{n_c^2 - n_{cl}^2} \qquad (10.3)$$

Rather than specifying θ_a, the quantity $n_0 \sin \theta_a$, known as the *numerical aperture* (NA),

$$NA = n_0 \sin \theta_a \qquad (10.4)$$

is usually specified for an optical fiber. This is most useful when a coupling medium other than air is used between the input light source and the fiber since the numerical aperture is independent of the source index.

EXAMPLE 10.1 The core and cladding indices are rather close in the typical case. Consider a plastic system with $n_c = 1.48$ and $n_{cl} = 1.46$. What are the numerical aperture and θ_a for input from air and from a coupling fluid with index 1.38?

Solution Using equation (10.3),

$$NA = \sqrt{(1.48)^2 - (1.46)^2} = 0.2425$$

and this is independent of the input medium. For an air interface this corresponds to a value of θ_a given by

$$\theta_a = \sin^{-1} 0.2425 = 14°$$

For an external index of 1.38, one has

$$\sin \theta_a = \frac{0.2425}{1.38}$$

using equation (10.4) and $\theta_a = 10.12°$. As one can see, the angles for the cone of acceptance are rather shallow and significant measures are required to couple the input light with the fiber.

All light falling within the numerical aperture will propagate in the fiber. Just as in a waveguide, however, only certain modes of propagation are supported by the fiber, and these *propagation modes* can be described by propagation at specific angles within the fiber. These angles, ζ, may be found approximately from the fact that the field across the waveguide must be a standing wave which vanishes at the walls, and thus

$$\sin \zeta = \frac{q\lambda}{2d} \qquad q = 0, 1, 2, \ldots \qquad (10.5)$$

where q specifies the mode, d is the core diameter, and λ is the wavelength of the radiation.

EXAMPLE   Find all the possible modes for the fiber in Example 10.1 if the core diameter is 10 μm,
10.2   the input radiation has a wavelength of 1.2 μm in the fiber, and the input medium is air.

Solution   Using equation (10.5),

$$\sin \zeta = \frac{q \times 1.2}{2 \times 10} = k \times 0.06$$

and the values of $\sin \zeta$ less than or equal to the numerical aperture of 0.2425 are

q	$\sin \zeta$	ζ
0	0.00	0.00°
1	0.06	3.44°
2	0.12	6.89°
3	0.18	10.34°
4	0.24	13.89°

so that only five modes will propagate under the given conditions.

As can be seen from the example, a limited number of modes exist, although in some instances it is possible to have thousands of modes propagating in the fiber. The velocity of propagation along the fiber is mode dependent, as will be seen. This gives rise to a dispersion, a spreading of the input signal which is mode dependent and which is called *modal dispersion*. This is avoided by a narrow-angle launch in fibers which are not limited to a single mode. A wide-beam launch or an off-angle launch into the fiber will result in modal dispersion and in some cases the launching of thousands of modes.

In the case of a multimode launch into a fiber, the distribution of modes varies along the fiber for some distance due to *mode coupling*, the transfer of energy between modes. Only after a propagation distance of several kilometers will a steady-state mode distribution be reached. Mode coupling occurs at bends in the fiber. Figure 10.4 illustrates this for the conversion of a low-order mode into a high-order mode and Figure 10.5 shows microbend mode coupling. Microbends are often artificially introduced at the front of short-run fiber systems to set up a steady-state mode distribution quickly. This is done by squeezing the fiber between two blocks with rough surfaces, mode mixing blocks.

If a pulse is input in a fiber with just two modes, one at angle ζ and a second axial mode with $\zeta = 0°$, there is a dispersion between these modes as shown in Figure 10.6. For a unit-length optical path, the equivalent path for the mode at ζ is $1/\cos \zeta$, and this results in a delay in the pulse along the angulated path given by

$$\Delta\tau = \frac{1}{v_1}(l_\zeta - l_0) = \frac{1}{v_1}\left(\frac{1}{\cos \zeta} - 1\right) \tag{10.6}$$

where v_1 is the velocity of propagation in the fiber. As one can see from Example 10.2, the angle ζ is usually small so that using the first two terms of the Taylor expansion for the cosine

$$\cos \zeta \approx 1 - \frac{1}{2}\zeta^2 \tag{10.7}$$

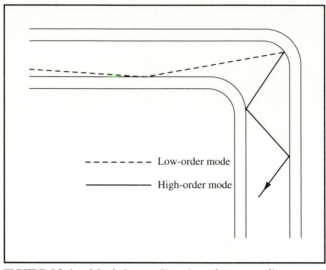

FIGURE 10.4.   Modal coupling in a large radius (macro) bend.

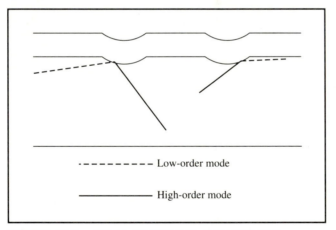

FIGURE 10.5.   Mode coupling in a microbend.

Since the numerical aperture is given by

$$NA = n_1 \sin \zeta \qquad (10.8)$$

with the small-angle approximation

$$\zeta \approx \frac{NA}{n_1} \qquad (10.9)$$

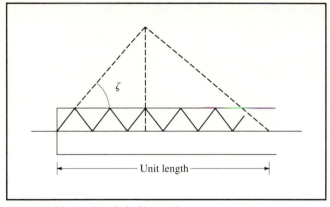

FIGURE 10.6. Modal dispersion.

Substituting these in equation (10.6) and using the definition of n_1 as c/v_1, one gets

$$\Delta\tau = \frac{n_1}{c}\left(\frac{1}{1 - \zeta^2/2} - 1\right) = \frac{n_1}{2c}\zeta^2 = \frac{(NA)^2}{2cn_1} \qquad (10.10)$$

The quantity $\Delta\tau$ is an important factor in fixing the pulse frequency at which one can communicate along the fiber. If $\Delta\tau$ for propagation along the full fiber length is equal to the pulse width assuming a uniform pulse width, then at the output of the fiber the pulse along the angulated path will reach the output at the trailing edge of the axial pulse and will result in a loss of information. As a rule one restricts ΔT to be

$$\Delta T \le 0.7T \qquad (10.11)$$

where T is the pulse width. This fixes the maximum bit rate and generally prevents collision of bits. From equation (10.10) one sees that the *fiber capacity*, a measure of the number of bits per second which can be sent down the fiber, obeys

$$\text{fiber capacity} \propto \frac{1}{(NA)^2} \qquad (10.12)$$

again illustrating the importance of the numerical aperture.

EXAMPLE 10.3

A fiber link of 10 km is to be established as part of a local area network. The numerical aperture of the fiber is 0.2425 for this fiber which has $n_c = 1.46$. What is the maximum frequency at which this network can safely be used?

Solution

For maximum frequency, safe use of the network must satisfy the equality in equation (10.11),

$$\Delta T = 0.7T$$

The delay per unit length is given by equation (10.10),

$$\Delta\tau = \frac{(0.2425)^2}{2 \times 3 \times 10^8 \times 1.46} = 6.7 \times 10^{11} \text{ s/m}$$

and

$$\Delta T = 6.7 \times 10^{-11} \text{ s/m} \times 10 \times 10^3 \text{ m} = 6.7 \times 10^{-7} \text{ s}$$

From this, $T = 9.6 \times 10^{-8}$, giving a safe frequency of approximately 10 MHz.

In addition to the modal dispersion, there is the *material dispersion* which one always has in a material medium as well as the *waveguide dispersion*. These result in a $\Delta\tau$ given by

$$\Delta\tau = \frac{1}{c}\frac{dn_1}{dk}\frac{1}{\cos\zeta}\Delta k + \frac{n_1}{c}\frac{d}{dk}\left(\frac{1}{\cos\zeta}\right)\Delta k \qquad (10.13)$$

where k is the wave number. The first term of (10.13) is the material dispersion, and the second term is the waveguide dispersion. As can be seen, these can be minimized by using a narrow-width laser source and a single-mode fiber.

Table 10.1 shows the importance of losses within the fiber and the impact of these losses on fiber systems is clear. This structural problem persists unless the fiber losses are minimized. Without this there is little possibility for other than short-run fiber optical systems. Figure 10.7 shows the attenuation characteristics of a silica fiber. There is an attenuation minimum at about 1.6 μm in the infrared region. The dashed lines show the theoretical limits which approximate a V shape. On the left the losses are due to Rayleigh scattering, which is caused by irregularities in the fiber of the order of 10^{-3} to 10^{-2} μm and which is proportional to λ^{-4}. The right arm of the V is due to absorption by the molecular components of the silica, an intrinsic material property.

The deviation from the theoretical curve is most apparent near the minimum. These deviations are due to impurities in the fiber material, principally Cr^{3+} and Fe^{2+} high on the left arm but most importantly, OH^-, near the minimum. The OH^- arises from the residual water in the fiber material and huge efforts are required to reduce the water content to less than one part per billion to eliminate these absorption bands.

It is clear that while fiber systems may be used in short-run applications, even with less than optimal attenuation, the use of optical fibers for long-run applications requires that amplifiers be placed at intervals along the fiber link. That is what has been done in the trans-Atlantic cable, for example, and not only are amplifiers included in the cable system but pulse reshaping circuitry is included to alleviate the problems of dispersion. One very interesting amplifier involves doping long-run optical cables with rare earth ions at intervals and using these ions as a laser medium at the frequency of the carrier. In such a case the amplifier acts to amplify coherently the signal coming down the cable.

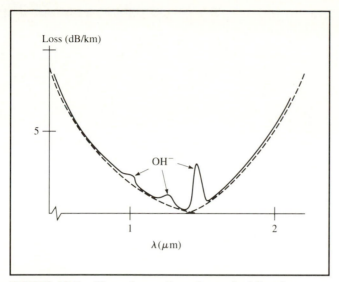

FIGURE 10.7. The attenuation characteristic of a silica fiber.

The rare earth doping of fibers has been done with neodymium, erbium, ytterbium, terbium, and praseodymium. The doped fibers themselves have impurity losses of about 3000 dB/km from the rare earth ions. When a second laser signal such as that from an argon ion laser is sent down the fiber along with the communications signal, the argon ion radiation serves as a pump for the rare earth ions, which then act as a laser and coherently amplify the signal on the fiber. The fiber is thin and the dopant ion sections are easily saturated by the pump radiation. The rare earth ions undergo laser transitions in the region of 1.5 μm, that is, at about the frequency at which the fiber losses are minimal. This technology is yet to be implemented but has great promise for use in long transmission line fiber systems. The attenuation minimum in most glass fibers is near the output wavelength of the GaAs diode laser so that it would seem that all factors fall into place for an ideal system. There are problems, however, and one of the most important of these is in joining fibers. The typical fiber has a diameter which is a fraction of a millimeter. Connecting two fibers becomes a difficult problem since any misalignment will not only result in scattering losses but will also produce a reflection of the signal back along the fiber itself. This becomes a particularly significant problem in the field, where the technician does not have access to a large automatic machine which can make accurate splices. The loss in decibels due simply to the offset of one fiber against a second is given by

$$L = 2.17 \frac{D}{\omega_0^2} \quad \text{dB} \tag{10.14}$$

where ω_0 is the spot size of the mode being propagated in μm and D is the offset displacement in μm. For a 1-μm offset and a spot size of 5 μm, the loss is about 0.1 dB, and one must budget at least that per splice in designing a fiber system.

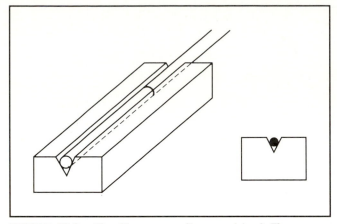

FIGURE 10.8. The V-groove alignment tool. The cleaved fibers are brought together and bonded in the aligning groove.

In addition to the radial misalignment, there is also an angular misalignment which results from the cleavage faces of the fibers to be joined not being normal to the centerline of the fiber. Occasionally there will be a gap left between two fibers in a splice, a longitudinal misalignment. For large multimode fibers with core diameter more than 50 μm, the target for splices is less than a 1-dB loss per splice while for the single-mode fibers with cores < 10 μm, the target is 0.1 dB, which requires alignment accuracy of 1 μm, a distance difficult even to measure.

Splices are made in three ways: welds, ferrule connectors, and lens couplers. In a weld the ends of the fibers are brought together and aligned and then either melted together or bonded with an optical cement. Alignment can be achieved by placing the fibers in a V-groove or other straightedge groove as shown for example in Figure 10.8. In a ferrule joint, the fibers are carefully cleaved to give flat faces at right angles to the centerline of the fiber, then clamped in a sheath. The two sheaths are spring loaded to maintain fiber abutment as in Figure 10.9. Finally, in a lens coupler, Figure 10.10, the beam is expanded, then recondensed onto the second fiber. Such systems have a relatively high insertion loss but are relatively rugged and can be used in military applications, for example.

Sources and Detectors

From Figure 10.7 it is evident that the optimal signal wavelength for the fiber in the figure would be about that of the GaAs diode laser, and to minimize the problems of material dispersion, the source for a fiber link is usually chosen to be a GaAs laser diode rather

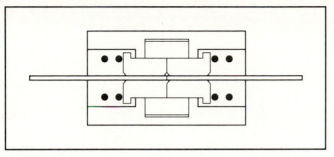

FIGURE 10.9. A ferrule connector showing the several sleeves used to ensure alignment. The system is spring loaded to keep the fibers tightly together.

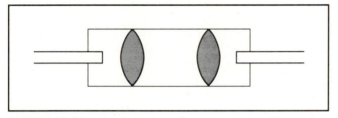

FIGURE 10.10. An imaging fiber connector. The output of the first fiber is imaged into the cone of acceptance of the second fiber.

than an LED. LEDs also are less efficient than lasers since they are wide-angle sources, and even when the fiber is inserted in the well of a Burrus-configured LED, the coupling efficiency is poor. LEDs also suffer by being more nonlinear than laser diodes when being driven by a signal.

Thus laser diodes are the device of choice as a source in most fiber optic systems. They have a highly linear power versus current region as can be seen in Figure 9.20. Their output is generally well confined with a half-power output angle of $\pm 10°$ as compared with $\pm 30°$ for an LED as shown in Figure 10.11. One important feature of the laser diode is the rise and fall time, which commonly is of the order of 0.5 GHz. The rise and fall times are measured with a photodetector, so that in this bandwidth the detector is factored in. Larger bandwidth can be achieved by operating the laser diode in a continuous wave (cw) fashion and modulating the input drive current.

The coupling of the output of the source to the fiber is called *launching*. The launch efficiency or coupling efficiency is defined by

$$\eta = \frac{P_F}{P_S} \tag{10.15}$$

where P_F is the power coupled into the fiber while P_S is the source output power. The coupling efficiency depends on a number of factors—the type of fiber as well as the source device—and may be supplemented by a number of aids used to improve this

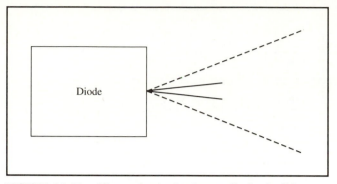

FIGURE 10.11. The output of a laser diode (solid lines) compared with that of an LED (dashed lines).

coupling. In practice many sources for fiber systems are provided with a short length of optical fiber called a *pigtail*. This reduces the optimization of the coupling to the preparation of a good splice, usually an easier process.

If one looks at the emitter-to-fiber coupling, there are a number of techniques which can be used to improve the coupling. When the fiber is simply abutted to the output device, there is a reflective loss at the fiber face given by equation (AIV.33) as

$$R = \left(\frac{n_f - 1}{n_f + 1}\right)^2 \tag{10.16}$$

which for a 1.5-index fiber gives an 11.1% reflective loss. This can be reduced by the use of a coupling fluid between the fiber and the source as shown in Figure 10.12. For a coupling fluid with index 1.3

$$R = \left(\frac{1.5 - 1.3}{1.5 + 1.3}\right)^2 = 0.0051 \tag{10.17}$$

and the coupling loss is reduced to 0.5%. One can find the loss in decibels by taking

$$P_F = (1 - R)P_S \tag{10.18}$$

The loss in decibels is

$$\text{loss} = 10 \log\left(\frac{P_F}{P_S}\right) = 10 \log(1 - R) \tag{10.19}$$

which gives a 0.022-dB loss in the presence of the coupling fluid while the loss is 0.51 dB with no coupling fluid.

There are a number of lens schemes which can be used to enhance the coupling of both LEDs and laser diodes to the fiber. Figure 10.13 shows three involving shaping the

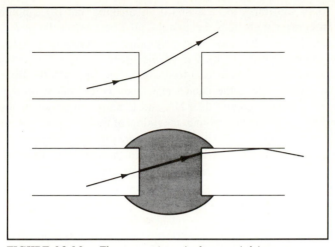

FIGURE 10.12. The use of an index-matching coupling fluid to increase coupling efficiency.

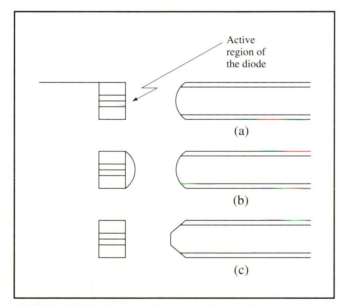

FIGURE 10.13. Source-fiber coupling schemes.

fiber face. In Figure 10.13a the face of the fiber is formed into a sphere to serve as a lens, while in Figure 10.13b, both the source and fiber have focusing faces. In Figure 10.13c, the end of the fiber is tapered to increase the angle of incidence between the fiber and the source radiation. The objective of any of these schemes is to ensure that the source fully illuminates the fiber core so that the maximum possible energy is launched.

Figure 10.14 illustrates the use of (a) an imaging and (b) a nonimaging sphere usually used in contact with both the source and fiber. While these techniques aid in coupling efficiency, they also have their own unique problems largely revolving around the small size of the spheres, which is of the order of the fiber diameter. With the nonimaging microsphere in Figure 10.14b it can be shown that if the fiber core radius is a and r_S is the source radius, the launch efficiency is dependent on the square of the numeric aperture if r_S/a is greater than NA and is unity otherwise. If the radius of the emitting area of the source is greater than the radius of the fiber core, one usually gets the best efficiency by butting the fiber to the active area of the source.

Many similar problems exist in coupling to the detector, although they are less severe since the detector diameter can always be made to exceed that of the fiber end.

For many years photodetectors have been based on the photoelectric effect. In the photoelectric effect as illustrated in Figure 10.15, a photon strikes a sensitized surface and causes an electron to be ejected from the surface. In the presence of the appropriate field, the electrons are swept away from the surface, causing a current to flow in the external circuit. There are two important issues to be considered. The *responsivity* of the detector is given by

$$\rho = \frac{I}{P} \tag{10.20}$$

where I is the output current from the detector and P is the light power falling on the detector. The *spectral sensitivity* is the wavelength variation of ρ. With the photoelectric effect, the spectral sensitivity is 0 for decreasing wavelength until a wavelength is reached

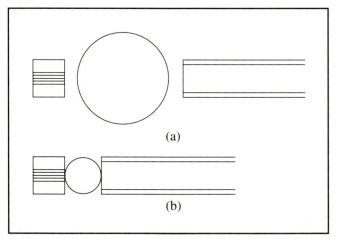

(a)

(b)

FIGURE 10.14.   Microsphere lens coupling schemes.
(a) Imaging. (b) Nonimaging.

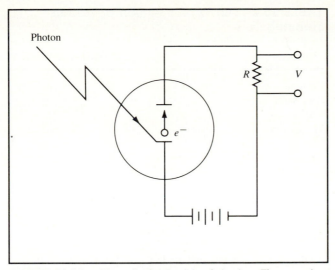

FIGURE 10.15. The photoelectric detector. The sensitized surface is in a vacuum and the enclosure has a window to admit the exciting radiation.

in which the energy of the photon is sufficient to cause an electron to be ejected from the surface. This energy, W_g, known as the *work function*, is given by

$$W_g = \frac{hc}{\lambda} \qquad (10.21)$$

and depends on the nature of the detector's surface. Table 10.2 lists the work function for several different materials. As can be seen from the table, cesium, with a work function of 1.9 eV, is the most responsive of the metal surfaces. The wavelength at which one has the onset of a photocurrent is given by

$$\lambda = \frac{1.24}{W_g} \qquad (10.22)$$

where W_g is in units of electron volts and λ is in microns. With cesium this cutoff value becomes 0.65 μm or 650 nm, a value in the far-visible red portion of the spectrum. Since cesium has the longest wavelength cutoff, it is clear that photoemissive devices such as these are of little use in fiber optic systems since they are totally insensitive in the region where fiber losses are minimal.

Even if a photon has an energy greater than the work function, it may not cause an electron to be ejected from the surface. This statistical process can be described by the *quantum efficiency* η of the emitter given by

$$\eta = \frac{\text{number of emitted electrons}}{\text{number of incident photons}} \qquad (10.23)$$

TABLE 10.2   The Work Function of Various Elements

Element	Work function (eV)
Cesium	1.90
Potassium	2.30
Sodium	2.75
Magnesium	3.66
Lead	4.25
Silver	4.62
Mercury	4.49
Cobalt	5.00
Gold	5.37

If W is the light energy striking the surface per unit time at frequency ν so that $W/h\nu$ is the number of incident photons per unit time, then the current flowing from the photocathode is given by

$$i_p = \eta \frac{eW}{h\nu} \tag{10.24}$$

and the *responsivity*, the current produced by radiation power W, is given by

$$\rho = \frac{i_p}{W} = \frac{\eta e}{h\nu} \tag{10.25}$$

The detector acts like a current source for the detector circuitry. The output voltage of the device shown in Figure 10.15 is then

$$V_p = \eta \frac{eWR}{h\nu} = \eta \frac{eW\lambda R}{hc} \tag{10.26}$$

where λ is the wavelength of the radiation and R is the load resistance.

The responsivity of a photoemissive detector can be increased by the use of a photomultiplier. In a photomultiplier as shown in Figure 10.16 the photoelectrons are made very energetic by a high accelerating voltage and collide with a number of intermediate electrodes known as *dynodes*. These energetic electrons collide with the surface of the dynode, causing several secondary electrons to be emitted at the surface, and these are then accelerated to the next dynode. This gives a multiplicative effect in that many electrons flow into the anode for each photoelectron initially generated. At each dynode, the number of secondary electrons emitted can be as high as 5 or 6. If one has a tube with 5 dynodes and a multiplication factor of 6, the gain is 6^5 or 7.78×10^3. Gains of 10^6 can be gotten with photomultipliers.

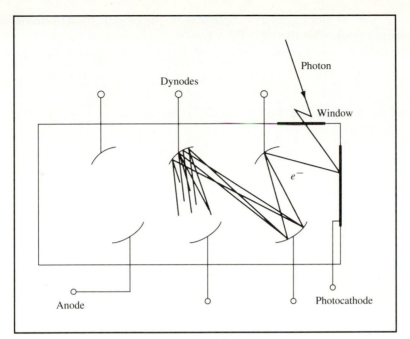

FIGURE 10.16.   A photomultiplier tube.

Even with photomultipliers the relatively large work function of the elements arises in part from the requirement that the photon-generated electron have sufficient energy to escape the surface of the material. To provide a detector which is useful in the near infrared, another approach is required. Photodiodes which act as sources were described in Chapter 9. There the diode was biased to drive carriers together in a recombination region so as to generate photons in the recombination process. If the diode is reverse biased so as to increase the gap between the bands in Figure 9.16, then electron–hole pairs will be quickly swept out of the recombination region with little opportunity for recombination. Electron–hole pairs can be created by photons falling on the active region of the diode. Such a device which uses photons to produce a current in a reverse-biased diode is a semiconductor photodiode.

Semiconductor photodiodes are typically *pin*-diodes (*p*-type, intrinsic, *n*-type) with the *i*-region being active. Pairs produced in the *n*- or *p*-regions will not be swept into the circuit quickly because of the small-field gradient in these regions and thus will often recombine. Charge carriers created in the central depletion layer will be accelerated across the depletion region and will be carried into the external circuit. In *pin*-photodiodes, which are the most commonly used detector in fiber optical systems, the *p*- and *n*-regions are kept thin so that there is a high probability that incoming photons will generate current-producing charge carriers.

The peak quantum efficiency of silicon and InGaAs *pin*-photodiodes is about 0.8. The peak response of the Si is at 800 nm, but that of the InGaAs is at 1.7 μm. The peak responsivity of InGaAs is 1.1 A/W as given by equation (10.25).

The rise time or speed of response of the *pin*-diode is fixed by the transit time for the charge carrier. The rise time for a typical *pin*-diode is of the order of 1 ns governed somewhat by the junction capacitance of the device. The device time constant is given by

$$\tau = 2.19RC \tag{10.27}$$

and the corresponding 3-dB bandwidth is given by

$$f_3 = \frac{1}{2\pi RC} \tag{10.28}$$

For a 2-ns rise time and a 5-pF capacitance, the bandwidth is 174 MHz.

The semiconductor photodiode parallels the vacuum photodiode with the exception of the work function. There is still a cutoff wavelength since the photon must have sufficient energy to create an electron–hole pair, but the energy required is considerably less than that required to remove an electron from a metallic surface. A device known as an avalanche photodiode parallels the photomultiplier again with a longer wavelength cutoff. The gain, however, does not approach that of the photomultiplier tube and is less than 200 as a rule. The gains are useful, however, in that they make these devices more sensitive than *pin*-diodes.

The avalanche diode uses the energy of the electron to create additional charge carriers when it collides with neutral atoms in the semiconductor. A single accelerated charge will produce several secondary charge carriers, and these in turn can produce additional carriers. The reverse bias in avalanche diodes may be as much as several hundred volts. The photocurrent is given by

$$i_p = M\eta\frac{eW}{h\nu} \tag{10.29}$$

where M is the device gain.

Communication Systems

Any communication system may be described by the basic elements shown in Figure 10.17. Fiber optic communication systems are no exception. The transmission channel is a fiber link. The principal uses for fiber optical systems to this point have been in telephone trunk lines and in local area networks for computer systems. At this time fiber links connecting the trunks to the home are beginning to be put into service. In the North American phone network the system takes the lowest rate signals, 64-Kbit/s voice signals, and multiplexes these through four levels to a signal rate of 274.176 Mbit/s. Recently an international standard, the SONET (synchronous optical network), has been established using somewhat different rates based on a line rate of 51.84 Mbit/s and building to 2488.32 Mbit/s.

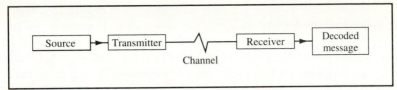

FIGURE 10.17. The block diagram for a general communication system.

There are many potential uses beyond phone networks. The potential for both narrow-band and broad-band uses in either analog or digital modes means that the fiber can carry phone, television, fax, telemetry, and so on, signals simultaneously. This suggests that ultimately fiber systems will completely revolutionize the way in which many communication functions are handled.

The limitation of a fiber communication link arising in fiber dispersion has already been considered in this chapter. Very-long-distance systems such as transoceanic fiber links use repeaters spaced about every 50 km to overcome the dispersion losses. These repeaters also contain circuitry to reshape the pulse. The fibers are typically single-mode operating in the 1.3–1.6-μm range where the dispersion is minimized. A copper line running in parallel with the optical fiber is used to provide power for the repeaters. The fibers themselves are so thin that many parallel lines can be housed in the volume usually occupied by a single copper wire.

The link can be run in any fashion in which copper lines are presently run, that is, undersea, underground, or aerially. An interesting installation of some links has placed them inside natural gas pipelines running between major cities. This has saved the expense of acquiring right-of-way for the cables and has meant that many kilometers of cable could be put in place each day compared with a few kilometers if trenching and cable burial were required. The volume of the gas line occupied by the cable is small enough to be neglected.

One of the most significant aspects of the design of fiber systems is the *power budget*. The power budget relates the receiver sensitivity for a given detector to the other parameters of the system such as fiber losses, splice losses, connector losses, and so on. The *power margin* is the power by which the signal at the receiver exceeds the receiver sensitivity. The error rate at the receiver is generally reduced as the power margin increases so that a highly functional system will have a large power margin.

Analog Systems

A model system using an analog transmission mode would be a closed-circuit television signal such as is used in security systems or in remote observation inside a dangerous environment as in a nuclear reactor. With television signals one needs a signal-to-noise ratio of about 46 dB to have a picture generally perceived as clear.

Systems such as these generally have short fiber runs of less than a kilometer and use either a step-index fiber or a gradient-index fiber. The source will be an LED in the 0.8–0.9-μm range and a *pin*-diode detector will function as the receiver. Performance improvement can be made with laser diodes or single-mode fibers, for example, but here the performance of the devices specified is sufficient. With 5 pF as the detector capacitance and a requirement of a 3-dB, 8-MHz cutoff, using equation (10.28) one finds a load resistance of 3979 Ω. One would use a load-resistor somewhat lower than this value to allow for degradations in the rest of the system. If the load used were to be 3600 Ω, the detector bandwidth would be 8.84 MHz.

The system will consist of the following elements:

(a) An LED with 1-mW output at 0.85 μm. It will be assumed to have a rise time of 12.5 ns and an emitter surface of 50 μm diameter.

(b) A step-index fiber with a 50-μm core, a loss of 5 dB/km, and a numerical aperture of 0.24.

(c) The *pin*-diode specified.

The power budget can be specified in dBm by referencing the 1-mW source. The receiver will be assumed to require −25.0 dBm so that the combined system losses must be less than 25.0 dB. These are, then,

Coupling loss $(NA)^2 = 0.0576$	12.40 dB
Reflection loss in input and output:	
2×0.2	0.40 dB
Connector losses	2.00 dB
	14.80 dB

The fiber losses can be 25.0 − 14.8 or 10.2 dB, and a 2-km fiber run will be within the design limits.

Digital Systems

Digital fiber optical communication systems involve the sending and receiving of light pulses, and the receiver recognizes "on" and "off" as the two states of the system. Here the discussion will be limited to a simple point-to-point link. The components which are assembled to create the link must be evaluated in terms of the signal-to-noise ratio at the receiver, that is, the energy budget, and in terms of the bit error rate. The latter can be improved by various coding schemes which help optimize the system.

Many point-to-point links are of such a length as to require repeaters to reshape and amplify the pulses. The need for repeaters can be assessed through the use of the length–bandwidth product *LB*, where *L* and *B* are the length and bandwidth, respectively, of an unamplified link. If the system is to operate with a bandwidth of B_0 and with a length L_0, then the number of repeaters required will be

$$N = \frac{L_0 B_0}{LB} \tag{10.30}$$

The power level at a distance d from the launch or from a repeater is given by the Beer–Lambert law

$$P(d) = P_0 \times 10^{-Ad/10} \tag{10.31}$$

where P_0 is the launch power and A is the fiber attenuation in decibels per kilometer. P_0 must be such that it is sufficient to drive the receiver, which requires a power P_R, the *receiver sensitivity*. The maximum length of a link is then

$$L_{max} = \frac{10}{A} \log_{10} \frac{P_0}{P_R} \tag{10.32}$$

Since P_0/P_R is often of the order of 10^5 in fiber systems, one can see that the critical factor is A. A change of an order of magnitude in P_0 will result in less than a 20% increase in L_{max} while a 25% increase in A will reduce L_{max} by 20%. It is notable that through-the-air systems such as radio have received power proportional to $1/R^2$, where R is the distance from source to receiver, so that an order of magnitude in transmission power triples the "link length."

The system used as an example here will consist of a laser diode source since these have a much better modulation speed and are better coupled to single-mode fibers. The launch power will then be about 1 mW, which is common for such systems. The receiver will be assumed to be an avalanche photodiode which requires about 10^3 photons/bit in the useful range of 0.8–1.6 μm, and the receiver sensitivity P_R is of the order of $10^{-13}B$ mW where B is the bit rate.

If P_i is the optical power incident at the receiver and is the quantum efficiency of the receiver, the error probability based on noise theory is given by

$$P_{err} = e^{-\eta P_i / h\nu B} \tag{10.33}$$

and for a $P_{err} \leq 10^{-9}$, $\eta P_i \geq 20$ photons/bit.

The bit rate is fixed by fiber dispersion as in equation (10.11). For a single-mode fiber, B can be of the order of 1 Gbit/s, while for a multimode fiber, B will be about 10 Mbit/s. The length–bandwidth product of 10^2 Gbit/s-km is well within the range of usable technology today. Figure 10.18 shows the length L versus B for a 1.6-μm wavelength.

The bandwidth required of a system is dependent on the *coding format*. Several of these are shown in Figure 10.19. In the return-to-zero (RZ) coding scheme with "0" represented by a low signal and "1" by a high signal, two transitions are required for each clock cycle. In the return-to-bias (RB) scheme, two transitions per clock cycle would also

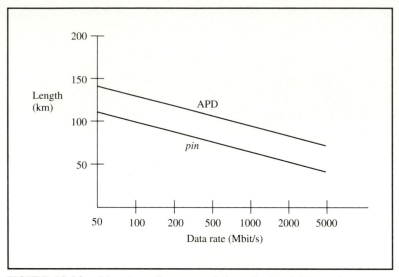

FIGURE 10.18.   Line length versus data rate for 1.55-μm laser input and an avalanche detector (ADP) or a *pin*-diode detector.

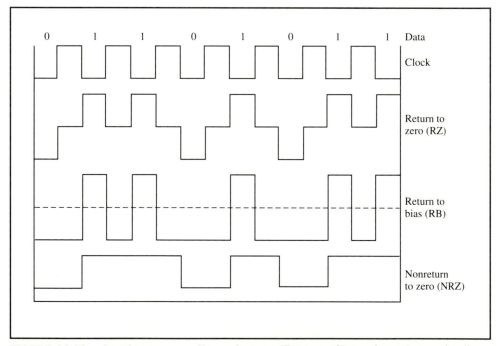

FIGURE 10.19.   Synchronous coding schemes. These coding schemes require the clock for synchronization.

be required in the worst-case pattern of 1111. . . . In the nonreturn-to-zero (NRZ) scheme, only a single transition is required for the worst-case signal 101010. . . .

Other coding schemes are prevalent. Those shown in Figure 10.19 are tied to the clock cycle. The codes shown in Figure 10.20 possess self-clocking capability and are known as *biphase codes*. Each of these codes requires at least one transition per bit period, and some of these have two in the worst case. Since there is a predictable transition during each bit period, these can serve to synchronize the receiver. Such codes are called *self-clocking codes*.

More complex coding schemes utilize block codes where a series of bits is treated as a set and encoded with an additional number of bits known as *check bits*. The bit error rate for various signals depends on the signal. A much lower bit error rate is acceptable for voice communications than for computer data exchange. The bit error rates of 10^9 (one error in 10^9 bits) for computer communications require the additional check bits.

The power budget of digital systems requires a number of steps. The signal at the receiver must first be shown to be large enough that one can separate high and low states. Only at that stage can the rise time be finally evaluated to be sure sufficient bandwidth is present to satisfy the specification. The link margin, M, is given by

$$M = P_0 - P_R - AL - A_c \qquad (10.34)$$

where A_c is the collected component loss due to connectors, splices, and so on. The powers may be peak power or average power, but one must be careful not to mix peak and average powers in the same calculation.

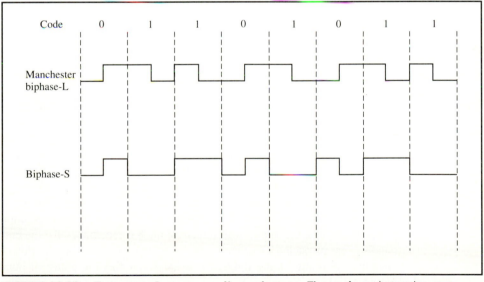

FIGURE 10.20. Two asynchronous coding schemes. These do not require any clock to synchronize the receiver.

**EXAMPLE
10.4**

A 50-μm core fiber with an attenuation of 0.4 dB/km is to transmit a 100-Mbit/s signal. The launched power will be 16 dBm, and the receiver sensitivity is -36 dBm. Three connectors with 1.5-dB loss each are part of the link. One is to allow a 7.5-dB loss for component aging or for link extension. What is the maximum allowable current link length? What extension would be possible if 4.5 dB of the 7.5-dB reserve is available for link extension?

Solution

Using equation (10.34),

$$7.5 = -16 - (-36) - 0.4 \times L - 3(1.5)$$
$$0.4L = 8$$

and the maximum allowable current link length would be 20 km. Link extension would require an additional connector (1.5 dB), leaving 3.0 dB for fiber losses, and ΔL would be $3.0/0.4 = 7.5$ km. The maximum extended link would then be no more than 27.5 km.

Fiber Cables—Coherent Bundles

Fibers rarely are used as single elements. Most commonly they are cabled together in much the same way that phone wires are cabled. The typical fiber cable will consist of a fiber bundle, which may be laid straight or helically around the central member, just like a copper cable. Most contain a strength element of steel or other strong fiber to support the cable and prevent excessive stress from being applied to the glass fibers themselves. Various other layers would include padding, water barriers, heat barriers, and external or internal armor.

One very interesting and widely used application of optical fiber bundles is as a flexible light source. A fiber bundle with or without a terminating lens can serve as a flexible light pipe, and light can be brought to areas otherwise unaccessible. The medical profession has made wide use of such devices to illuminate various body cavities such as the stomach or the lungs.

Once the illumination is available in otherwise unaccessible areas, it is necessary to view these areas and here again fiber systems are of great use. If one can form an image on the end of a fiber bundle, the individual fibers will transmit the illumination to the other end of the bundle. What has been done is to make the bundle *coherent*, that is, to set the bundle in such a fashion that individual fibers retain their relative positions; a fiber at the 8 o'clock edge of the bundle will remain at the 8 o'clock edge for the entire length of the bundle. The physician can then view the other end of the bundle outside the body and examine the lining of the lungs, for example. A flexible coherent bundle is shown in Figure 10.21.

Other possibilities exist with coherent bundles. The photograph in Figure 10.21 shows two of these. If the bundle is twisted as it is by 180° in the central large bundle in

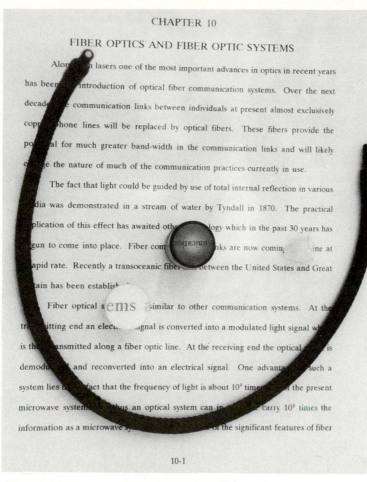

FIGURE 10.21 in the image reads:

CHAPTER 10

FIBER OPTICS AND FIBER OPTIC SYSTEMS

Along with lasers one of the most important advances in optics in recent years has been the introduction of optical fiber communication systems. Over the next decade the communication links between individuals at present almost exclusively copper phone lines will be replaced by optical fibers. These fibers provide the potential for much greater band-width in the communication links and will likely change the nature of much of the communication practices currently in use.

The fact that light could be guided by use of total internal reflection in various media was demonstrated in a stream of water by Tyndall in 1870. The practical application of this effect has awaited other technology which in the past 30 years has begun to come into place. Fiber communication links are now coming on line at a rapid rate. Recently a transoceanic fiber link between the United States and Great Britain has been established.

Fiber optical systems similar to other communication systems. At the transmitting end an electrical signal is converted into a modulated light signal which is then transmitted along a fiber optic line. At the receiving end the optical is demodulated and reconverted into an electrical signal. One advantage of such a system lies in the fact that the frequency of light is about 10^4 times that of the present microwave system, thus an optical system can in principle carry 10^4 times the information as a microwave system. One of the significant features of fiber

10-1

FIGURE 10.21. A coherent fiber bundle, an image rotator, and a fiber taper.

the figure, the image is rotated by that amount. Also, one sees a side view and a top view of a tapered bundle in the figure. Such a bundle can magnify the object with which it is in contact. The magnification due to the taper shown in the photo is 2×.

PROBLEMS

10.1. An optical fiber made of plastic with $n = 1.41$ with no cladding is to be used as a light guide. What are the angle of the cone of acceptance and the numerical aperture, assuming a launch from air? if the launch is from a matching fluid with $n = 1.43$?

10.2. The fiber in Problem 10.1 has an 80-μm diameter. Make a table listing several of the modes of propagation for this fiber for launch from the matching fluid with 1.2-μm light. What is the maximum frequency at which this fiber can be reasonable used for a 10-km link?

10.3. Estimate the material dispersion and the waveguide dispersion for the fiber used in the previous problems. How do these compare with the results found in Problem 10.2?

10.4. Assume that a system can tolerate a 0.4-dB loss at each splice. If the mean offset is 2 μm, how large a spot size would could be used?

10.5. A laser diode outputs into a pigtail 1-mW power. Assume that for test purposes the pigtail is accurately butted against the fiber but is not spliced. There is then an air interface between the pigtail and the fiber. How much power actually is launched along the fiber in this test setup? How might one improve this? Assume the pigtail and fiber have $n = 1.5$.

10.6. The launch in a fiber system is to be made using a fiber whose end is spherical. The fiber has an index of 1.54 and the cladding index is 1.51. The core diameter of the fiber is 80 μm. If the laser diode source is assumed to produce parallel light, how will the spherical end affect the launch?

10.7. Plot the work function versus cutoff wavelength for wavelengths from 2 to 0.4 μm.

10.8. Radiation at 500 nm falls on a photocathode with a quantum efficiency of 0.8. How much radiation power will give the minimum detectable 1 pW? How many photons/s?

10.9. The time constant for a *pin*-diode is found to be 2 ns. If the capacitance of the diode is 5 pF, what resistance will it have to have for a 3-dB bandwidth of 174 MHz?

10.10. What are the worst-case data (the data with the maximum number of transitions per cycle) for the coding schemes in Figure 10.19?

10.11. If a fiber has an attenuation coefficient $A = 0.2$ dB/km and $P_0/P_R = 10^4$, what is the maximum length of run permitted for such a system?

10.12. The link in Problem 10.11 is to be put in place between two islands separated by 750 km. Assume fixed bandwidth. How many repeaters will be required? Assuming a 1.5-dB connector loss in each link and a 1-dB aging reserve in each link, how many repeaters will be required in this case?

10.13. A fiber link of 200 km is to use a cable with a 0.2-dB/km attenuation. If no repeaters are to be used, what must the receiver sensitivity be for a 1-mW launched signal? Assuming that the receiver is 60% efficient and that 1.6-μm light with 50 Mbit/s is used, what is the error probability?

10.14. A fiber system is to run along the coast of Florida for a distance of 600 miles (1000 km). The 50-μm fiber has an attenuation of 0.2 dB/km at the 1.55-μm wavelength where the system is operated. Assume each connector has a loss of 0.5 dB and the receiver sensitivity ratio is 10^5. If taps are made every 50 km, reducing the signal by 3 dB at each of them in addition to a connector loss at the tap, set up a power budget for this system.

10.15. Conical fibers such as those in the tapers in Figure 10.21 appear brighter when viewed from the wide end with the narrow end against the page and darker when reversed. Discuss why this is the case.

10.16. The rotator in Figure 10.21 appears dark toward the edges and seems to function only centrally. Why?

CHAPTER 11

Optics as a Linear System, Fourier Optics, Optical Computing Devices

Throughout the many branches of engineering, the theory of linear systems is frequently applied as a method of analysis. In this regard, optics is no exception. The application of linear system theory in optics has produced some very exciting and interesting results. For example, much of the thrust toward an optical computer comes from the application of linear system theory, usually termed *Fourier optics,* as a result of the extensive use of Fourier transform techniques in the analysis of these linear systems. This chapter will explore the way in which linear system theory is applied to optical systems, and this will be illustrated with a number of examples.

Linear Systems

Linear systems are characterized by the linear nature of the action of the system on its input. Figure 11.1 illustrates a general "black box" linear system S acting on an input $f(x)$. The output $g(x)$ is given by

$$g(x) = \mathscr{L}f(x) \tag{11.1}$$

where $\mathscr{L}$ is a mathematical operator which characterizes the system acting on the input function $f(x)$. The linearity of the system implies that

$$\mathscr{L}\{A_1 f_1(x) + A_2 f_2(x)\} = A_1 \mathscr{L}f_1(x) + A_2 \mathscr{L}f_2(x) \tag{11.2}$$

Equation (11.2) is simply a formalized statement of the principle of superposition.

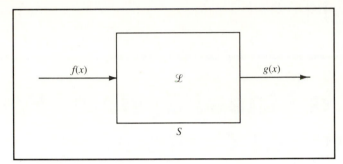

FIGURE 11.1. A "black box" linear system.

There are additional properties of linear systems which are also of importance. *Shift invariance* of the input is defined by

$$\mathcal{L}\{f(x - x_0)\} = g(x - x_0) \tag{11.3}$$

which simply states that the position of the input has no bearing on the behavior of the system. This restriction is often quite reasonable over a small, finite range but rarely is strictly true over an unrestricted range. Optical systems with this property are called *isoplanatic systems* and such systems are often designated as LSI—linear, shift-invariant— systems. In optics one more often speaks of an *isoplanatic patch*, a region where the system is isoplanatic even if the entire optical system is not.

The Impulse Reponse Function

The response of a system to a delta function impulse, $\delta(x)$, is of paramount importance for LSI systems as a result of equation (11.2).[1] If the response of the system to a unit impulse is known, then the output of the system can be found easily. The impulse response $h(x; x_0)$ is defined by

$$h(x; x_0) = \mathcal{L}\{\delta(x - x_0)\} \tag{11.4}$$

for a linear system whose operator is $\mathcal{L}$. For an impulse at the origin

$$\mathcal{L}\{\delta(x)\} = h(x; 0) \tag{11.5}$$

and as a result of the assumed shift invariance of the system, one can write

$$h(x; x_0) = h(x - x_0; 0) = h(x - x_0) \tag{11.6}$$

---

[1] The delta function as well as a short treatment of the Fourier transformation are included in Appendix II.

It follows from (11.2) and (11.6) that an LSI system can be completely characterized in terms of its unit impulse response.

Consider now an input to an LSI system with a unit impulse response $h(x - x_0)$. This input $f(x)$ will be defined for a range of values of x, but from the properties of the delta function, the output can be given by

$$g(x) = \int_{-\infty}^{+\infty} h(x - x_0)f(x_0)\ dx_0 \tag{11.7}$$

but (11.7) is simply the *convolution* of $h(x)$ and $f(x)$ so (11.7) can also be written

$$g(x) = h(x) * f(x) \tag{11.8}$$

using the convolution operator (AIII.21).

In optical systems one deals typically not with one-dimensional but with two-dimensional input and output. In this case

$$h(x,y) = h(x - x_0, y - y_0) \tag{11.9}$$

and the output $g(x,y)$ is given by

$$g(x,y) = h(x,y) * f(x,y) \tag{11.10}$$

where the two-dimensional convolution (11.10) is defined as

$$g(x,y) = \int_{-\infty}^{+\infty} \int h(x - x_0, y - y_0)f(x_0,y_0)\ dx_0\ dy_0 \tag{11.11}$$

The convolution theorem (AIII.24) in two dimensions parallels that in one dimension and is written in the frequency domain as

$$G(f_X, f_Y) = H(f_X, f_Y)F(f_X, f_Y) \tag{11.12}$$

and employs the two-dimensional Fourier transforms defined by

$$\mathscr{F}\{f(x,y)\} = F(f_X, f_Y) \int_{-\infty}^{+\infty} \int f(x,y)e^{-j2\pi(f_X x + f_Y y)}\ dx\ dy \tag{11.13}$$

The task now becomes one of defining the frequencies, which here are *spatial frequencies* with units of reciprocal length. The spatial frequencies were introduced in (AII.37) and were defined to be the quantities

$$f_X = \frac{x_0}{\lambda z} \qquad f_Y = \frac{y_0}{\lambda z} \tag{11.14}$$

Each of these has the units of reciprocal length, just as frequency in the time domain has units of reciprocal time. Figure 11.2 illustrates a two-dimensional object with periodicity along the x axis. This object has a spatial frequency of 9 cycles/mm. Complex two-dimensional objects may be built up from two-dimensional Fourier transforms just as a one-dimensional time-domain signal can be built up from frequency components of the Fourier transform of a time-domain signal. This can be done on an optical bench quite easily, as will be seen shortly.

It is shown in Appendix II that the optical amplitude in the diffraction field is a Fourier transformation of the optical disturbance in the aperture of the diffraction screen. As an illustration, consider a one-dimensional line source parallel to the y axis at infinity and a screen with two slits parallel to the y axis in the $z = 0$ plane at $\pm d_0$. The source and the two slits will be taken as delta functions. Using convolution one gets

$$\int_{-\infty}^{+\infty} \delta(\alpha)[\delta(d_0 - \alpha) + \delta(-d_0 - \alpha)]d\alpha = \delta(d_0) + \delta(-d_0) \qquad (11.15)$$

a two-slit source whose Fourier transform is

$$\mathscr{F}\{\delta(d_0) + \delta(-d_0)\} = 2 \cos 2\pi f_x d_0 \qquad (11.16)$$

The frequency f_x is given by (11.14). This is just Young's experiment, and one finds intensity maxima at

$$n\pi = \frac{2\pi x d_0}{\lambda z} \qquad (11.17)$$

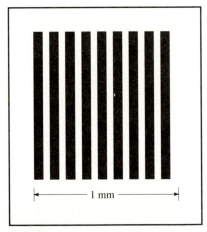

FIGURE 11.2. A two-dimensional object periodic in one dimension with a spatial frequency of 9 cycles/mm.

for the slit separation $d = 2d_0$ as was found in Chapter 7 using a different approach. This same result can be gotten from equation (11.12), and that is left as an exercise.

An optical system defined in terms of $h(x,y)$ or $H(f_X,f_Y)$ can be viewed as a filter acting on the input wave form. This is best approached using the frequency-domain representation $H(f_X,f_Y)$. Most generally, $H(f_X,f_Y)$ will be a complex quantity and as such may be expressed as

$$H(f_X,f_Y) = A(f_X,f_Y)e^{-j\phi(f_X,f_Y)} \tag{11.18}$$

that is, in terms of its real modulus A and its real phase ϕ. Systems such as the two slits described are amplitude filters; they are purely real. A lens, on the other hand, alters the phase of the light wave as well as the amplitude through its aperture. A lens can be thought of in terms of a *pupil function* defined by its aperture and a phase function which alters the phase of the incident wave.

The presence of a lens in the pupil introduces a phase change in the wave traversing the pupil and imaging by a lens may be treated in terms of this phase change. Consider the planoconvex lens in Figure 11.3. A biconvex lens can be treated as two back-to-back plano lenses, so that this simpler case will be sufficient to establish the principle. The alteration in phase of a wave traversing a lens is due to the slowing of the wave in the medium of the lens where the refractive index is higher than in the surround. The thickness of the lens on the optic axis will be taken as Δ_0. At a height x above the optic axis, Δ_0 is divided into two parts: that made up of lens, Δ, and the free-space portion, Δ_1:

$$\Delta_1 = R_1 - \sqrt{R_1^2 - x^2 - y^2} \tag{11.19}$$

where $(x^2 + y^2)$ is the square of the normal distance of P from the optic axis.

$$\Delta = \Delta_0 - \Delta_1 \tag{11.20}$$

for the part in the lens, Δ. Δ_1 can also be written

$$\Delta_1 = R_1 - R_1\left(1 - \frac{x^2 + y^2}{R_1^2}\right)^{1/2} \tag{11.21}$$

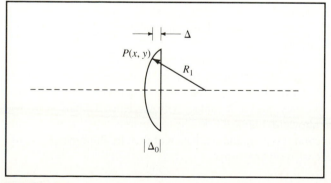

FIGURE 11.3.

and in the paraxial approximation where $x,y \ll R_1$, the square root may be taken as

$$\left(1 - \frac{x^2 + y^2}{R_1^2}\right)^{1/2} = 1 - \frac{x^2 + y^2}{2R_1^2} \tag{11.22}$$

using the first term of the Taylor expansion and Δ becomes

$$\Delta = \Delta_0 - \frac{x^2 + y^2}{2R_1} \tag{11.23}$$

In the case where neither surface of the lens is plano,

$$\Delta = \Delta_0 - \frac{x^2 + y^2}{2}\left(\frac{1}{R_1} - \frac{1}{R_2}\right) \tag{11.24}$$

where Δ_0 remains the center thickness of the lens and R_2 is the radius of curvature of the second surface.

The phase change in a wave passing through the lens is now a function of x and y, that is, of the distance from the optic axis. This is given by

$$\phi(x,y) = \frac{2\pi}{\lambda}n\Delta(x,y) + \frac{2\pi}{\lambda}[\Delta_0 - \Delta(x,y)] \tag{11.25}$$

and, noting that $2\pi/\lambda$ is the wave number k, the exponential of the phase becomes

$$e^{jkn\Delta_0}e^{-jk(n-1)(x^2+y^2/2)(1/R_1-1/R_2)} \tag{11.26}$$

which, using the lensmaker's equation (2.22) yields

$$e^{jk\Delta_0}e^{-j(k/2f)(x^2+y^2)} \tag{11.27}$$

where f is the focal length of the lens. The first term is simply a constant phase delay and can be dropped, while the second term remains as the phase alteration as a function of position within the lens pupil.

The lens then functions as a pure phase filter, and this property leads to an important use for lenses beyond that of image formation. Initially, the assumption will be made that the lens aperture is infinite, and thus the lens will not limit the light which will be passing through a transparency placed in contact with the lens. A plane wave will illuminate the transparency as shown in Figure 11.4. The transparency will have its transmittance represented by a function $t(x_1,y_1)$, and the lens will then be illuminated by $At(x_1,y_1)$, where A is the amplitude of the plane wave:

$$U = At(x_1,y_1) \tag{11.28}$$

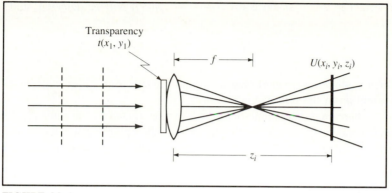

FIGURE 11.4.

After passing through the lens the optical disturbance has a phase distribution imposed upon it by the lens (11.27), and one has

$$U' = At(x_1, y_1)e^{-j(k/2f)(x_1^2 + y_1^2)} \qquad (11.29)$$

where the prime (') is used to indicate that this amplitude distribution is behind the lens. The constant phase term in (11.27), which is of no interest, has been dropped.

The quantity U' is the field distribution in an aperture formed by the lens–transparency combination. Using the Fresnel expression (AII.35), one gets

$$U(x_0, y_0) = \frac{e^{2jkz}}{j\lambda z} e^{j(k/2z)(x_0^2 + y_0^2)} \int_{-\infty}^{+\infty}\!\!\int U'(x_1, y_1)e^{j(k/2z)(x_1^2 + y_1^2)}e^{-j2\pi(x_1 f_x + y_1 f_y)}\, dx_1\, dy_1 \quad (11.30)$$

for the optical disturbance in the plane at z. If one now substitutes for U' using (11.29) and lets $z = f$, that is, observes the distribution of radiation in the focal plane of the lens, (11.30) becomes

$$U(x_0, y_0) = A\frac{e^{jkf}}{j\lambda f}e^{j(k/2f)(x_0^2 + y_0^2)} \int_{-\infty}^{+\infty}\!\!\int t(x, y)e^{-j2\pi(x_1 f_x + y_1 f_y)}\, dx_1\, dy_1 \quad (11.31)$$

The distribution of radiation in the focal plane of the lens depends on the Fourier transform of the transparency

$$U(x_0, y_0) = A\frac{e^{jkf}}{j\lambda f}e^{j(k/2f)(x_0^2 + y_0^2)}\, \mathcal{F}\{t(x_1, y_1)\} \qquad (11.32)$$

At the focal plane, then, the optical disturbance is proportional to the Fourier transform of the transparency but contains a quadratic phase factor. This quadratic phase factor is not important in this case since it is $I = U^*U$, which will be detected, and

$$I = U^*U = \frac{A^2}{\lambda^2 f^2} \, |\mathscr{F}\{t(x_1, y_1)\}|^2 \tag{11.33}$$

I then is direction proportional to the power spectrum $|\mathscr{F}\{x_1, y_1\}|^2$ of the transmittance distribution in the transparency.

There are other geometries which can be treated by the same approach. Consider a transparency not in contact with the lens but rather placed at a distance d before the lens as shown in Figure 11.5. In this case with $t(x, y)$ as the transmittance function of the transparency, the Fourier transform of the distribution falling on the lens is given by the Fresnel diffraction pattern of the transparency,

$$e^{-j\pi\lambda d_1(f_X^2 + f_Y^2)} \, \mathscr{F}\{t(x, y)\} \tag{11.34}$$

After passing through the lens the wave acquires the phase term given by (11.27) and the optical disturbance at the focal plane of the lens is given by

$$U(x_0, y_0; f) = \frac{e^{j(k/2f)(1 - d_1/f)(x_0^2 + y_0^2)}}{j\lambda f} \mathscr{F}\{t(x, y)\} \tag{11.35}$$

The optical field in the focal plane is again found to be proportional to the Fourier transform of the transparency. This result differs from that in (11.32) by one significant factor. If d_1 is taken as f, that is, if the transparency is in the front focal plane of the lens, then the phase term

$$e^{j(k/2f)(1 - d_1/f)(x_0^2 + y_0^2)}$$

which causes the phase to vary over the focal plane becomes constant. One then has a precise Fourier transform relationship. *Equation (11.35) is the keystone expression for optical signal processing.*

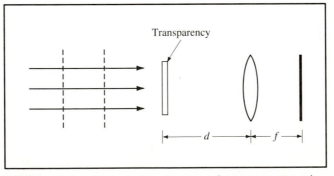

FIGURE 11.5a.   The geometry for a transparency set d before the transforming lens.

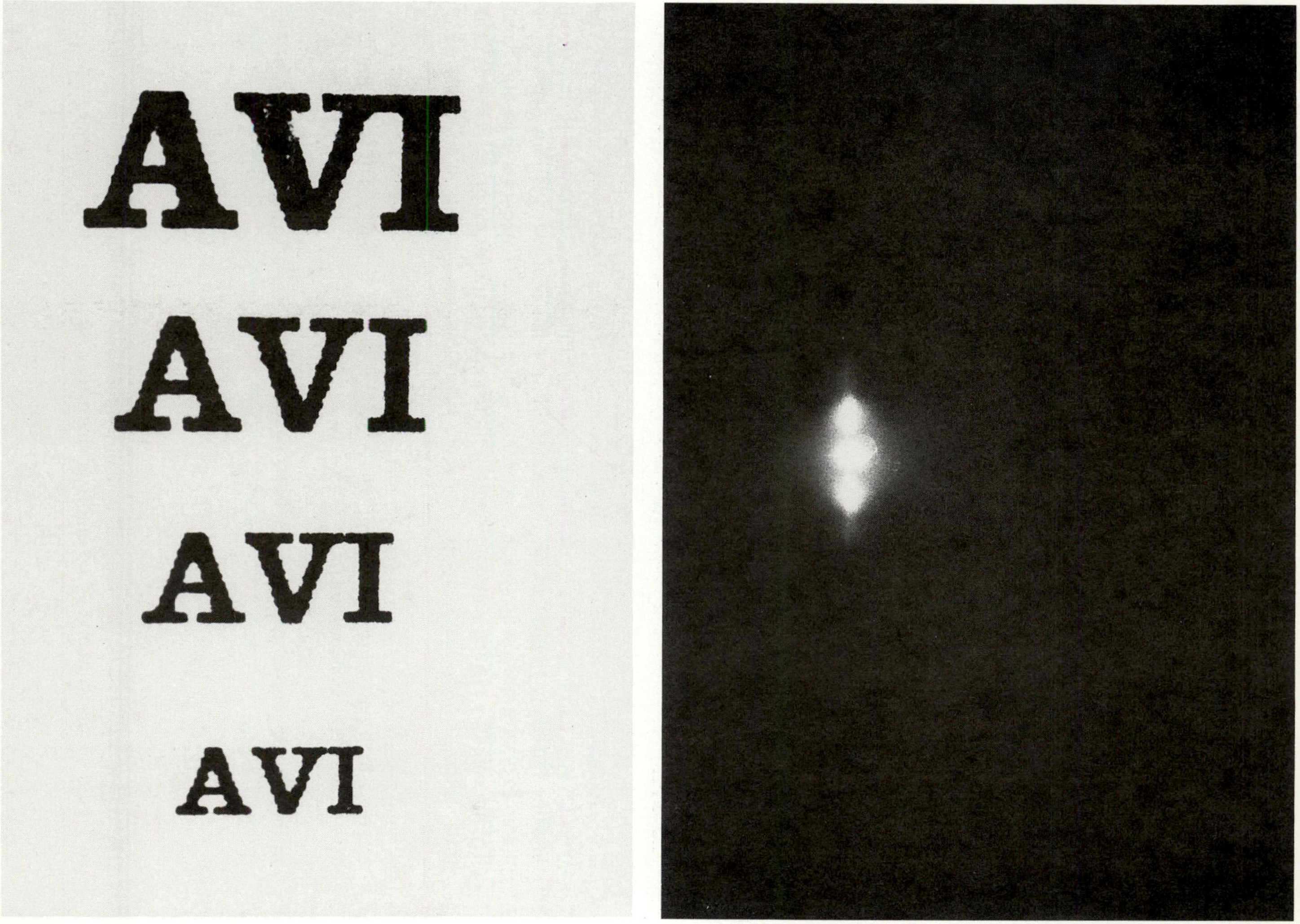

(b)

(c)

FIGURE 11.5b,c.   An object and its Fourier transform using the illustrated geometry.

EXAMPLE 11.1

A transparency with a vertical sinusoidal distribution of wavelength 0.1 mm is in the object-side focal plane of a lens of focal length 10 cm and is illuminated with light at 632.8 nm. What does one observe in the image-side focal plane of the lens?

Solution

The object has a spatial frequency of 10 lp/mm.[2] The Fourier transform of such an object is given in Table AII.1 as

$$\mathcal{F}\{\cos(2\pi\nu_0 x)\} = \tfrac{1}{2}[\delta(\nu - \nu_0) + \delta(\nu + \nu_0)]$$

With the given frequency, one will find two vertical lines in the Fourier transform plane corresponding to this frequency. To locate them, one uses (11.14),

$$f_X = \frac{x_0}{\lambda z}$$

where z is the focal length. One has

$$x_0 = 10 \text{ lp/mm} \times 632.8 \times 10^{-6} \text{ mm} \times 100 \quad \text{mm}$$

$$= 0.6328 \quad \text{mm}$$

and at ± 0.6238 mm about the axis of the system, a vertical bright line will appear in the Fourier transform plane.

Example 11.1 is one of the simplest examples of an optical Fourier transform. The object was an extremely simple one. Most objects are much more complex and contain not just one but many frequencies in two dimensions. What one does get, however, is the ability to place a filter in this plane. For example, high spatial frequencies in an object can be eliminated by placing a circular aperture in the Fourier transform plane. The cutoff in these filters is very sharp, unlike electronic filters, as can be seen in Example 11.2.

EXAMPLE 11.2

Design a high-pass filter with a cutoff at 100 lp/mm for use with a lens of focal length of 10 cm and light at 623.8 nm.

Solution

In the Fourier transform plane one finds, using (11.14), for a spatial frequency of 100 lp/mm,

$$r_0 = f\lambda z = 100 \text{ lp/mm} \times 623.8 \times 10^{-6} \text{ mm} \times 100 \quad \text{mm}$$

$$= 6.238 \quad \text{mm}$$

The region with $r < 6.238$ mm will contain all the low frequencies and a filter as shown in Figure 11.6b will serve. The cutoff is sharp with no rolloff to either side.

---

[2] lp = line pairs.

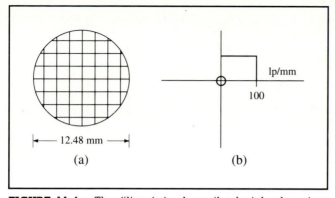

FIGURE 11.6. The filter (a) where the hatched region represents a hole in an opaque plate. The passband is shown in (b).

In a system such as that in Example 11.2, the filtered image will appear in the far field, that is, at infinity. Often one needs to record the filtered image on film, for example. In such a case one uses a pair of lenses and the experimental arrangement shown in Figure 11.7. Most often L_2 is "paired" with L_1 so that the filtered image has a magnification of 1; it is possible to alter the size of the filtered image by using an L_2 of different focal length than L_1, in which case the experimental arrangement is as shown in Figure 11.8.

It is significant that the Fourier transform of an object is found in the focal plane since, if the object is moved from the optic axis in its plane, the Fourier transform remains on the axis. This can be seen in Figure 11.9. As a result it is possible to detect a given object anywhere in the object field with a matched filter centered on the transform plane. If one wants to determine if the letter "A" is present in the object field, one need only place a filter mask which is matched to the transform of "A." Other objects may also give some response but the maximum response will come from A.

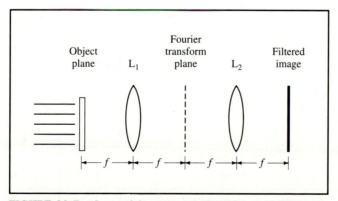

FIGURE 11.7. L_1 and L_2 are a pair of lenses of equal focal length used to present the filtered image in the near field.

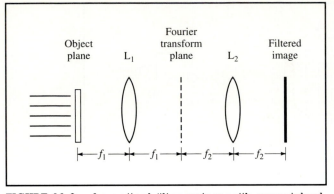

FIGURE 11.8.   An optical filter system with unmatched lenses.

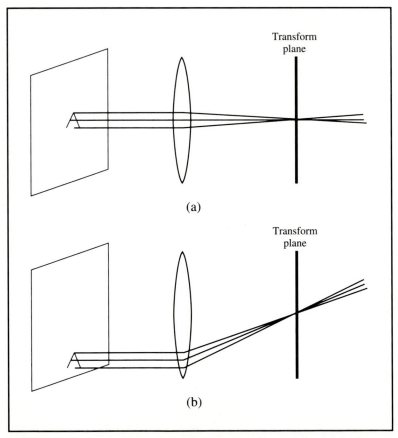

FIGURE 11.9.   The effect of shifting the position of the object on its spectrum in the Fourier transform plane.

The Fourier transform of A is given by

$$\mathcal{F}\{A\} = |F(f_X, f_Y)|e^{j\phi(f_X, f_Y)} \tag{11.36}$$

The matched filter is one which has a transmittance given by

$$t_{\text{fil}} = |F(f_X, f_Y)|e^{-j\phi(f_X, f_Y)} \tag{11.37}$$

The output beyond the mask is the product

$$|F(f_X, f_Y)|e^{j\phi(f_X, f_Y)} \times |F(f_X, f_Y)|e^{-j\phi(f_X, f_Y)} = |F(f_X, f_Y)|^2 \tag{11.38}$$

This is a parallel beam as a result of the filter having stripped out the phase term, as can be seen in equation (11.38). This can be detected in the conjugate plane. What is found there is a spot image known as the *correlation peak*. This is illustrated in Figure 11.10.

The system just shown can be used for *pattern recognition* and represents one of the simplest applications of the optical Fourier transformation. Filters are commonly made using holographic techniques.[3] Unfortunately, this technique is extremely sensitive to the rotation of the object as well as to the relative size of the object and thus has not found as extensive a use as one might expect.

Incoherent Light

Throughout the earlier sections of this chapter the implicit assumption has been made that the light is coherent, so that any phase difference between a pair of points such as a and b in Figure 11.11 remains fixed over time. This requires a monochromatic source, that is, a laser. While lasers are quite common, light is often produced in a hot, incoherent, white source such as a mercury–halogen lamp with a colored filter to limit the light to a narrow spectral range. Using a pinhole and a very narrow band-pass filter as in Figure 11.12, the light can be made partially coherent, but one really needs a laser to approach full coherence.

The output intensity of an incoherently illuminated object is given by

$$I(x_0, y_0) = \langle U^*(x_0, y_0)U(x_0, y_0)\rangle \tag{11.39}$$

where U is the optical disturbance and the brackets represent the time-averaged value. With a source given by $U(x, y, t)$ and a transfer function $h(x, y)$, then

$$I(x_0, y_0) = \langle (U^* * h^*) \times (U * h)\rangle \tag{11.40}$$

---

[3] Holograms will be discussed in Chapter 12.

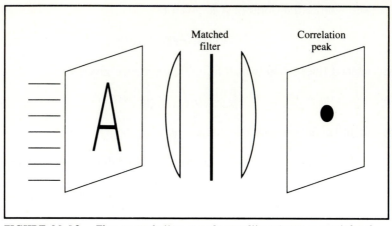

FIGURE 11.10.   The correlation peak resulting from a matched filter in the Fourier transform plane.

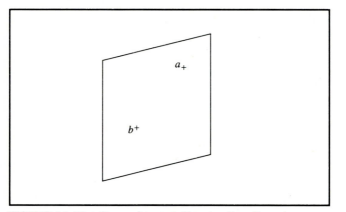

FIGURE 11.11.   For coherent illumination the phase difference $\phi_b - \phi_a$ is independent of time.

If the system transfer function $h(x,y,t)$ is time-independent as is usually the case, then

$$I(x_0,y_0) = h * h * \langle U * U \rangle \tag{11.41}$$

or

$$I(x_0,y_0) = |h^2| * I(x_1,y_1) \tag{11.42}$$

where x_1 and y_1 are the coordinates in the plane of the aperture. Thus the difference between a coherent system and an incoherent system is the fact that the convolution in the incoherent system is between the *input intensity* and the *modulus of the system transfer*

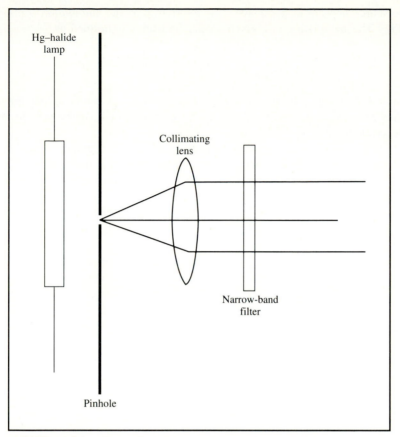

Hg–halide lamp

Collimating lens

Narrow-band filter

Pinhole

FIGURE 11.12. The production of narrow-band light using an incoherent source.

function, while for the coherent case, the convolution is between the transfer function and the input disturbance and the output intensity is then found as the squared modulus of that convolution product.

The Modulation Transfer Function

The quantity $|h|^2$ which appears in equation (11.42) is not only useful in evaluating the output of a system, but it can also be used to evaluate the quality of a lens. $|h|^2$ is the modulus of the transfer function and thus is a real quantity depending on the spatial frequencies f_X and f_Y. In effect it measures the way a given spatial frequency appearing in the source appears in the image. It is referred to as the *modulation transfer function* or MTF. The modulation at a given spatial frequency is defined as

$$M = \frac{I_{\max} - I_{\min}}{I_{\max} + I_{\min}} \tag{11.43}$$

where $I_{\max}$ and $I_{\min}$ are the maximum and minimum intensities at that spatial frequency. What M is, of course, is a measure of the contrast which can be present in the image at the given frequency.

The MTF has a maximum value of 1 and falls to 0 at some generally high spatial frequency, the *cutoff frequency*. Figure 11.13 shows the theoretical MTF for a perfect lens. As can be seen in the figure, the MTF has its maximum value at zero frequency and is not quite linear. The perfect system which would have such an MTF is called a *diffraction-limited system* since the only contribution to the MTF is from diffraction; that is, there are no aberrations due to the lens, which would have the effect of depressing this MTF curve.

Real lenses are not usually diffraction limited; tolerance errors in their construction and aberrations present in their design serve to alter the MTF curve. The MTF can easily serve as a test of lens quality and is used in the quality control of lenses.

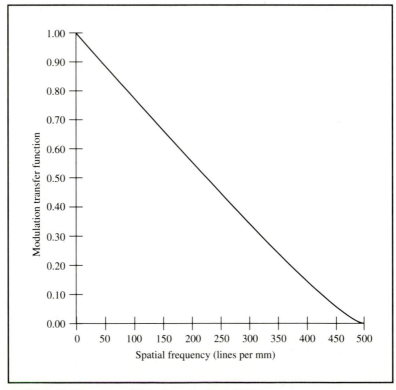

FIGURE 11.13. The diffraction-limited MTF.

Figure 11.14 is a photograph of a test target which has been widely used to evaluate the MTF of lenses. The target consists of a series of bars at varying spatial frequency, and these are divided into groups with each group having a series of group elements. One such group element is shown in Figure 11.15. This element represents one spatial frequency. When this element is imaged through the lens the contrast of the target group is initially 1 as in Figure 11.16a, but the finite aperture of the lens as well as any imaging defects in the lens itself due to aberrations and the like give the image the appearance of Figure 11.16b. The contrast has diminished. One might think of this in terms of the shadow of a straightedge, as in Figure 8.1. When the edges are close, the edge diffraction reduces the contrast in the region between them. In the presence of lens aberrations, this effect is compounded.

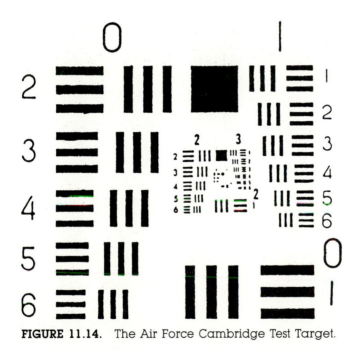

FIGURE 11.14. The Air Force Cambridge Test Target.

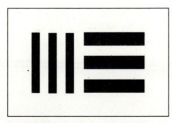

FIGURE 11.15. One element of the test target.

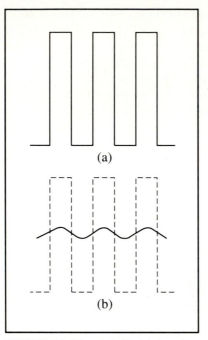

FIGURE 11.16. (a) The intensity distribution of one element of the test target. (b) The modulated intensity.

The MTF of a perfect lens, that is, one which is diffraction limited, is given by

$$\text{MTF} = \frac{2}{\pi}(\phi - \cos\phi\,\sin\phi) \tag{11.44}$$

where

$$\phi = \cos^{-1}\left(\frac{\lambda f}{2\text{NA}}\right) \tag{11.45}$$

Here f is the spatial frequency, NA is the numerical aperture of the lens, and λ is the wavelength of the light used. The design MTF can be calculated by taking the modulus of the Fourier transform of the image distribution of a fan of rays in the image plane.

By comparing the measured MTF with the theoretical value, one can make judgments of the quality of production lenses or of the success of a prototype in meeting the design expectation. This is a common practice today. Figure 11.17 represents three curves taken on production lenses. As one can see, they are near to the theoretical value and would be considered acceptable for most uses. They were taken using one of several commercially available instruments for this purpose.

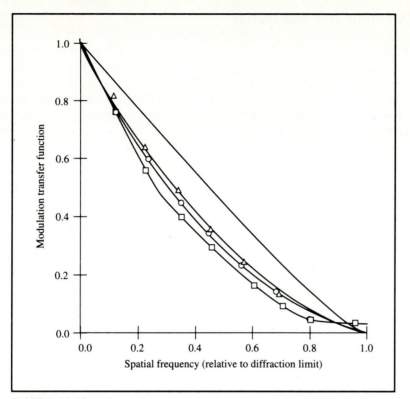

FIGURE 11.17. The MTF values of three production lenses compared with the theoretical expectation for these lenses.

Optical Computers

There are a number of features of light which make the goal of an optical computer significant. One is speed. The velocity of an optical signal is close to the limiting velocity for signal propagation of c, the free-space velocity. Another is the lack of cross-talk. Optical signals can cross in space and not be affected by one another since they are charge neutral. There are additional reasons for seeking such a computer, but the two just mentioned are sufficient in themselves.

Earlier in this chapter the Fourier transforming properties of a lens were illustrated. This is an example of a dedicated computer, one which is devoted to a single task. In what follows several other examples of dedicated functions will be illustrated. No effort will be made to cover the optical implementation of all arithmetic operations, but only a few distinctive examples will be discussed. The questions regarding a general multipurpose optical computer, while currently being addressed by many research workers, will be limited to one interesting example of an optical switch.

Addition and Subtraction

Consider two input functions $f(x,y)$ and $g(x,y)$ in the form of transparencies where one wants to form the sum $f + g$. Figure 11.18 illustrates how this can be done. A collimated monochromatic light source is used and its beam divided into two paths in the form of a Mach–Zender interferometer. The phase is adjusted with the input transparencies removed so that the light in the two arms adds. The transparencies are then inserted one in each arm as shown in the figure, and at the viewing screen one finds their sum.

Subtraction can be accomplished with the same arrangement. One simply uses the phase shifter and aligns the system so as to cause destructive interference in the output plane. When the transparencies $f(x,y)$ and $g(x,y)$ are inserted, the phase change in one arm results in the subtraction of the images.

Multiplication

This is the simplest function to implement since it requires no device. When light passes through two transparencies which are sandwiched together, if I_0 is the incident intensity, then the output is given by

$$I = g(x,y)[f(x,y)I_0] = [f(x,y)g(x,y)]I_0 \tag{11.46}$$

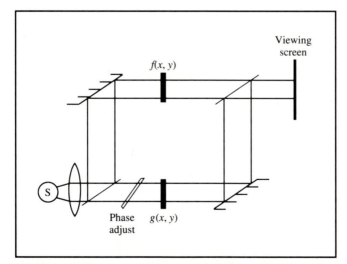

FIGURE 11.18. An addition–subtraction optical module. The phase is adjusted by a plate similar to that used in interferometers.

Integration

The transparency $f(x,y)$ represents a two-dimensional function. If this is illuminated by a parallel light beam and the output is brought to a focus, the intensity of the light at the focal spot is a measure of the double integral over the area of the transparency. A single integral along any axis can be gotten by using a cylindrical lens rather than a spherical lens.

Differentiation

There are two approaches to differentiation. In the first one uses the definition of a derivative

$$\frac{\partial f(x,y)}{\partial x} = \lim_{\Delta \to 0} \frac{f(x + \Delta x, y) - f(x,y)}{\Delta x} \tag{11.47}$$

and simply subtracts one image from a slightly displaced copy of itself.

The second approach makes use of the Fourier transforming properties of a lens. Given a function $g(x,y)$, its derivative can be found as

$$\frac{\partial g(x,y)}{\partial x} = \mathscr{F}^{-1}\ \mathscr{F}\left[\frac{\partial g(x,y)}{\partial x}\right] \tag{11.48}$$

and

$$\mathscr{F}\left[\frac{\partial g(x,y)}{\partial x}\right] = j2\pi f_X G(f_X, f_Y) \tag{11.49}$$

What is required is a filter given by $j2\pi f_X$ placed in the Fourier plane of the lens in Figure 11.19. The problem of the negative values of f_X is solved by using two filters, one as in

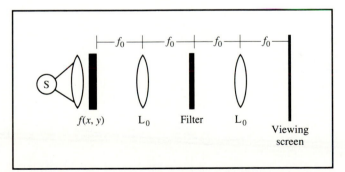

FIGURE 11.19. The device for multiplication using a Fourier filter.

Figure 11.20a which is $|2\pi f_X|$ and the second a phase filter as in Figure 11.20b, sand-wiched together. The derivative (11.47) is seen in the viewing plane when $g(x,y)$ is the input.

Matrix Operations

One of the more interesting of the optical computational devices is that dealing with vector transformations and matrix products. Figure 11.21 illustrates the principle. An LED array represents the input which may be intensity coded. A cylindrical lens is used to image this

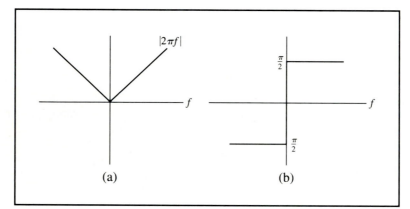

FIGURE 11.20.   The make-up of the multiplication filter.

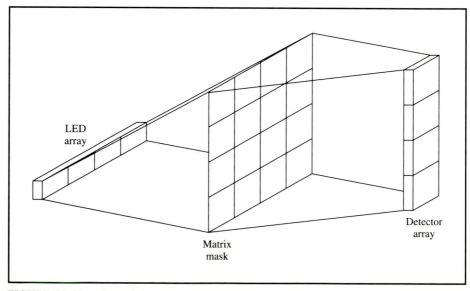

FIGURE 11.21.   A matrix array processor.

onto a matrix mask, the cylindrical lens producing a line image across the matrix array. On the output side of the matrix a second cylindrical lens at right angles to the first focuses the output from the matrix onto a photodetector array.

Through a sequencing of the values in the LED array, one can achieve matrix–matrix multiplication.

The Transparency

The photographic transparency has been used as an input example without really being defined in detail. Typically one does not simply use a photographic slide since thickness variations lead to unacceptable phase retardation. To avoid this problem, the transparency is usually enclosed in a liquid gate consisting of two glass plates, with the transparency enclosed between them in an index-matching fluid as in Figure 11.22. This has the effect of removing the phase variations across the film.

An interesting alternative to a photographic transparency is the use of a liquid crystal television. Such systems are readily available in black and white for less than \$100. When the screen is removed from its case and inserted in a liquid gate, it can be programmed to function as an input device or in the filter plane. The principal disadvantage is the fact that the contrast ratio is low, as is its resolution. Nonetheless, liquid crystal television offers some intriguing possibilities for optical processing.

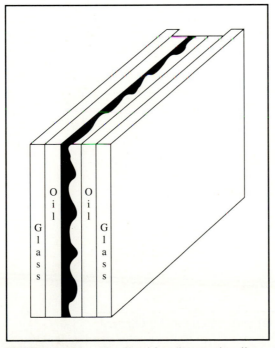

FIGURE 11.22.   The liquid gate construction.

An Optical Switch

While the general multipurpose optical computer will not be discussed, one example of a critical element, the on–off switch, will be given. Multipurpose computer architectures generally make such switches key elements in their implementation.

A *surface acoustic wave* or SAW device is a crystal transducer, that is, a device in which input electronic waveforms are converted into surface acoustic waves. Light which passes through the SAW device perpendicular to the waves will be phase modulated by the waves. This light can then be filtered in a Fourier transforming optical cell as shown in Figure 11.23. The zero-order stop prevents light from passing through the system when no wave is present on the crystal, that is, in the off state. Such a filter is often termed a dc filter since it has the effect of not only filtering the unexcited system but also removing the constant term from any Fourier transform. When a wave is present in the transducer, the central maximum of the transform is filtered, but the modulated light passes through the filter and is detected, the on state.

If the SAW device is used in reflection as in Figure 11.24, it can also serve as a frequency analyzer. The input wave is on the surface of the device, and it will scatter the incident beam as a function of the frequency of the input acoustic wave. The surface appears as a multiple aperture grating. With a suitable set of detectors, the input signal can be frequency analyzed. The device is similar to a grating, but the scattered waves are reflected rather than transmitted.

Liquid crystal systems and polarization devices are two more of the many ways in which on–off switches can be implemented.

These examples of optical computing devices are merely a handful of the many which have been proposed. Many new examples appear in the literature every month. Within a few years it is expected that many of these will be used commercially.

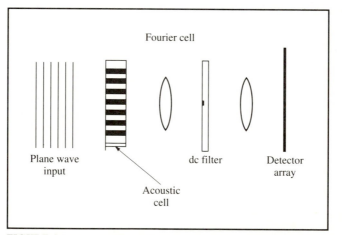

FIGURE 11.23. A Fourier cell switch for a multipurpose optical computer.

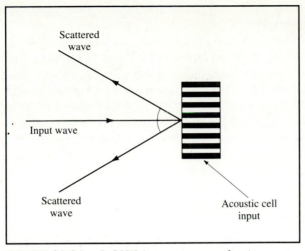

FIGURE 11.24. A SAW frequency analyzer.

PROBLEMS

11.1. The square-wave rect(x) is defined as 1 for $|x| < \frac{1}{2}$, as $\frac{1}{2}$ for $x = \frac{1}{2}$, and zero otherwise. Find the convolution

$$\text{rect}(x) * \text{rect}(x)$$

11.2. The unit step function step(x) is defined as 0 for $x < 0$, as $\frac{1}{2}$ for $x = 0$, and 1 otherwise. Find the convolutions

$$\text{step}(x) * \text{rect}(x)$$

$$\text{step}(x) * \text{step}(x)$$

11.3. Show that the phase change across a concavoplane lens is consistent with equation (11.23) when proper account is taken of the sign of R.

11.4. Discuss a transparency placed between the Fourier transforming lens and its focal plane. What does one find in the focal (transform) plane?

11.5. A spark chamber photograph of a nuclear particle decay has horizontal bars which somewhat obscure the data in the photograph. Design a filter to remove the bars. Assume that the photo is in the anterior focal plane of a lens with focal length 10 cm and that the spatial frequency of the bars in the photo is 12/mm. ($\lambda = 500$ nm.)

11.6. If one photographs a TV picture, one finds that the horizontal raster lines are visible in the photo. The photo is a 35-mm image, so that the 525 raster lines are distributed over 24 mm of the slide. Design a filter using the lens in Problem 11.5 and compare your result here with that from Problem 11.5.

11.7. The "ideal" MTF in equation (11.44) results in a nearly linear plot as in Figure 11.13. Investigate the linearity of this expression by expanding the trigonometric terms.

11.8. Figure 11.21 represent a vector–matrix multiplier. Make a similar drawing using a two-element vector and a 2×2 matrix, assign values to each of the elements, and demonstrate that the output vector is indeed the product. Write the equivalent mathematical equation.

11.9. Investigate the dc filter in Figure 11.23. What is its size for a Fourier transforming cell with lenses having 10-cm focal length? Given a practical dc spot, what will be the low-frequency cutoff?

11.10. A SAW device has a surface wave propagation velocity of 4.5×10^3 m/s. It is interrogated by a normally incident laser at 632.8 nm. The output is detected on an array extending from $5°$ to $40°$ at $1°$ intervals. If the input is at 480 MHz, where is the output detected? What range of frequencies can be detected by this device?

CHAPTER 12

Image Recording and Holography

Until the 19th century, images were recorded only by artists. There were no techniques available to record an image of an event other than the skill and the eye of the artist sketching the scene. In the early 19th century, photography was introduced, so that by the middle of that century, the important figures and events of the time were recorded. There is a considerable body of photographs of the American Civil War, for example, made by Matthew Brady using the Daguerreotype process, named for J. L. M. Daguerre, who along with N. Niepce and Sir John Herschel among others, launched the science of photography.

The real growth of photography came at the end of the 19th century when in 1888 George Eastman introduced a portable roll-film camera, the No. 1 Kodak, making photography available to amateurs as well as professionals. In this century electronic recording has provided yet another alternative to the artist's eye. Both photography and electronic image recording will be treated in this chapter.

Photographic Film

In Chapter 4 the lenses used to produce an image on a photographic plate were discussed. Here the attention is directed to the photographic film, the medium upon which this image falls and is recorded. Photographic films are classified as either color or black and white, the latter being used almost exclusively until recently and still the film of choice for many scientific purposes, including, for example, the recording of medical x rays.

The photographic process involves a series of steps: the exposure of the sensitive film in a camera or other system, the development of the film to yield a *negative,* processing of the negative to fix the image into a stable, semipermanent record, and finally, if required, secondary exposure and development of a positive image *print*. Often the negative is a suitable form for the image, and the latter step is not required, as is the case with the image in Figure 12.1.

The film itself consists of a sensitive layer of finely divided crystalline grains of a silver halide material uniformly distributed in a gelatin binder. The silver halide–gelatin is coated onto a backing material such as acetate, paper, or glass, which serves as a mechanical support. The silver halide–gelatin is universally referred to as the *emulsion* by photographers, but it is a dispersion, not a true emulsion. The selection of the silver halide is a factor in determining the film's sensitivity. Silver bromide with small amounts of the iodide is used in the most sensitive films, while silver chloride with traces of the bromide is used in the least sensitive or *slowest* emulsions.

The backing also allows the formation of uniform layers of the emulsion. In the most delicate processes, the backing is a glass plate, while for most commercial applications a flexible backing is provided. The original flexible sheet which was used was

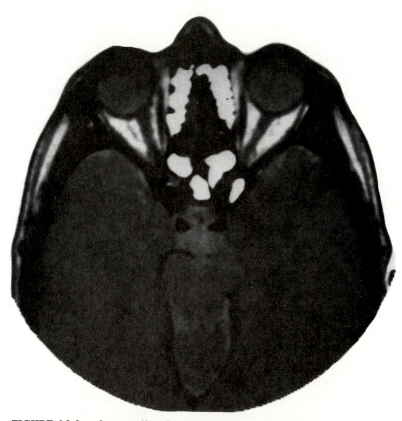

FIGURE 12.1. A negative image from a computerized tomography scan showing the cross section of the head. The negative provides the information, and it is not necessary in this case to produce a positive print.

cellulose nitrate. This material was not only extremely flammable but somewhat unstable, and, as a result, many old photographs and early motion pictures have been lost through the decay of this base material. Today cellulose triacetate is the common base material, eliminating many of the problems with the nitrate. The positive photographic print is usually made onto a paper base. The backing is usually of the order of 0.1 mm thick, but thicker bases are used for x rays or motion pictures where the negative images are subject to extensive handling. A supercoat is applied over the relatively soft emulsion to prevent scratching, and an undercoat to the emulsion is often used to strengthen the binding to the backing.

An additional layer is used as a backing on film. This is called an *antihalation layer*. If light is reflected from the back of the film and passes up through the emulsion a second time, the images of bright objects are seen to be surrounded by halos. The antihalation layer attenuates the light which has passed through the base and prevents this problem. The antihalation layer is generally an anticurl layer as well, to help maintain the flatness of the film. Figure 12.2 shows the cross section of a typical film.

Color films are similarly constructed but use three separate emulsions sensitive to red, green, and blue light. It is interesting to note that it was Maxwell who first conceived of and described this color process.

Film Characteristics

Exposure of photographic film to light releases photoelectrons within the fine grains of silver halide dispersed in the gelatin. A photoelectron from the halide ion neutralizes a silver ion and forms an atom of silver metal. The metal atom within the crystalline structure of the silver halide distorts the crystal structure and forms a center of sensitivity, causing other photoelectrons to be attracted to that region and neutralizing other silver ions there. The reaction can be expressed by

$$Br^- + h\nu \rightarrow Br + e^-$$

$$Ag^+ + e^- \rightarrow Ag$$

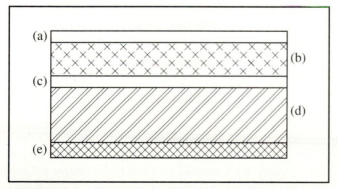

FIGURE 12.2. The cross section of a photographic film: (a) supercoat; (b) emulsion; (c) binding substrate; (d) backing; (e) antihalation–anticurl layer.

The exposed film, now containing crystals with silver metal, is termed the *latent image*. It is necessary to convert this to a developed negative by chemically processing the film. The film, maintained under dark conditions, is immersed in a developer solution which contains a reducing agent that converts those crystals containing silver metal clusters to silver while leaving the silver halide crystals without any silver metal centers unchanged. The film is then "stopped" and "fixed." The stop bath halts the development. Fixation consists of thoroughly washing the film with a solution, usually sodium thiosulfate, which complexes with the otherwise insoluble silver halides and makes them soluble. The complexed silver halide is then washed out of the film. At this point further exposure of the film to light will not cause any further darkening of the film. Times and temperature for the development are critical, and these are carefully specified by each manufacturer of both film and developing solutions.

This process of exposure and development is not linear. It depends not only on the developing chemistry but upon which silver halide or what mixture of silver halides makes up the emulsion, as well as the average size of the halide crystals. The Beer–Lambert law describes the transmission characteristic of a developed film,

$$\tau(x,y) = Ae^{-d(x,y)} \tag{12.1}$$

where d depends on the quantity of metallic silver present at that point in the negative. This can also be expressed as

$$d = \log\left(\frac{1}{\tau}\right) + \log A = D + \log A \tag{12.2}$$

where the quantity $\log(1/\tau)$ is usually called the *optical density D*.

Two pioneers in photographic science, Ferdinand Hurter and Vero C. Driffield, characterized film by plotting D versus the log of the exposure. This plot has come to be known as the *H-D curve* and is used to characterize various films. Figure 12.3 is one example of an H-D curve. The curve consists of three sections: section I, the toe; section II, the linear portion; and section III, the shoulder.

The density does not go to zero in region I. The quantity D_{min} is called the *fog* or the *chemical fog* and represents the density of the unexposed regions of the film. The fog density depends on both the developing process and the emulsion. In very precise measurements D_{min} is often subtracted from the experimental D value to establish a more precise measurement of the exposure radiation.

The shoulder of the curve in region III represents the maximum density which can be gotten regardless of the amount of exposure. Further exposure once D_{max} has been reached actually results in a decrease in D (beyond the scale in Figure 12.3), a process known as solarization or bleaching.

The linear portion of the H-D curve is the critical feature. Its slope $\Delta D/\Delta(\log E)$ is called the *gradient G*. The gradient is constant only in the linear portion of the curve, where it is commonly called the *gamma* of the film, Γ. Γ contains information only about the straight-line portion of the curve. The larger the value of Γ, the faster the film will be; that is, less exposure will be necessary to reach a given D. The intersection of the extrapo-

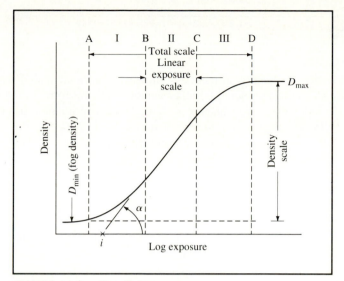

FIGURE 12.3. The H-D curve.

lation with the abscissa is called the *inertia, i*. Using this the film can be characterized by the linear expression

$$D = \Gamma \log E - \log i \qquad (12.3)$$

The sensitivity of film is measured in terms of what constitutes the exposure necessary to produce a given density D. Two standards are commonly used and are marked on most commercially available film. These are the European (DIN) system and the American (ASA) standard. Both accomplish the same thing in that they give some measure of the slope, Γ, of the H-D curve. The DIN is derived from the exposure necessary to achieve a density 0.1 above the fog.

With a film one wants to reproduce as density differences the luminance differences of the object being photographed. This suggests that one use some measure of the density across the range of the film and this has been adopted as the ASA standard. An exposure E_0 which produces a density 0.1 above the fog is first measured. One then find the point where the gradient is 0.3Γ, a point known as the *speed point*. The density is then determined from E_0 at an exposure 1.3 log units greater than E_0. The development time is chosen so that ΔD, the density difference, is about 0.8. The average gradient between the two points is then about 0.62, while Γ is usually slightly higher. The speed is then

$$S = \frac{0.8}{E_0} \qquad (12.4)$$

with E_0 in m-candle-s.

Electronic Imaging

Film represents but one method of recording images. Images are also directly taken and transmitted as TV signals or stored in either analog or digital form on magnetic tape or computer disks using electronic imaging system. Electronic imaging has the additional advantage of allowing one to alter contrast and color balance within the image. Indeed today with certain electronic image enhancement technologies one can literally ''see in the dark'' electronically.

Vidicon

The vidicon is a generic example of the classical vacuum tubes used until recently in TV systems. Figure 12.4 represents a typical tube. The camera lens focuses the image onto a high-resistance photoconductive layer deposited on the back surface of its entrance window. This layer serves as its photosensitive element. A scanning beam of electrons is swept across this element in a series of horizontal lines, the *raster pattern*. Wherever the photoconductive layer has been exposed, when that position is struck by the interrogating electron beam, the surface-to-cathode potential, initially at 30 to 50 V, changes. The current flow in the external circuit then acts as a measure of the exposure of the photosensitive layer.

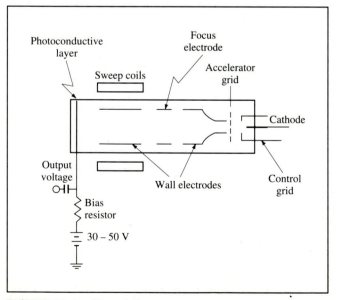

FIGURE 12.4.   The vidicon.

In the United States the image frame is swept 29.97 times a second, and the raster consists of 525 horizontal lines. This results in a 4.2-MHz bandwidth. The European standard is 25 frames per second but with 625 raster lines, and this gives a higher-resolution picture, but the bandwidth required is increased to 5.0 to 6.0 MHz.

A more sensitive tube, the orthicon, has the photoconductive layer set back in the tube from the photocathode. This is shown in Figure 12.5. An accelerator grid is set between the photocathode and the photoconductive plate. Photoelectrons generated by the incident light are accelerated away from the photocathode and strike the photoconductive plate at higher energy than the light provides. The orthicon can operate in much lower light levels than the vidicon.

A serious disadvantage with these vacuum tube images is the size of the associated electronics. An electron beam must be generated to read the image, and it must be swept across the tube, usually by a combination of electric and magnetic fields. Cameras using this technology are typically both heavy and bulky.

The Charge-Coupled Device

An alternative to the vidicon involves the use of a discrete array of detectors in either one or two dimensions. If the one-dimensional linear array is used, the image is swept by one of the many optical scanning devices currently available. In these solid-state devices, known as *charge-coupled devices,* or CCDs, the problem is to read out sequentially the output of each detector, when even a simple array of 512 $\times$ 512 elements contains 2.62 $\times$ 10^5 elements.

The CCD is based on the metal-oxide-semiconductor (MOS) capacitor. In the CCD the substrate base *p*-type silicon material has deposited on it a layer of insulating silicon

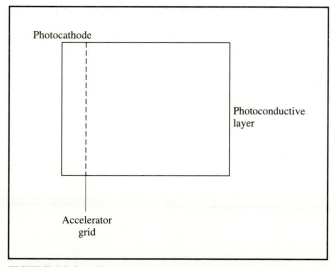

FIGURE 12.5.   The front end of the orthicon tube.

dioxide, SiO$_2$, topped with a metal electrode as in Figure 12.6. The metal electrode, called the *gate*, is biased positively with respect to the silicon substrate. Photon-generated electron–hole pairs will separate, and the electrons will be attracted to the surface and will accumulate under the gate. The accumulated charge will be proportional to the light flux falling on the device between reads.

At readout time, the trapped charge is read sequentially along a line of detectors. This is done by passing the charge down the line from detector to detector. There are a number of ways in which this can be done. Fundamentally, the gate potentials are supplied from three voltage lines, L$_1$, L$_2$, and L$_3$, which are connected to each third electrode in sequence as in Figure 12.7. L$_1$ is kept at a positive potential, while L$_2$ and L$_3$ are maintained at ground. Photoelectrons will be accumulated under L$_1$ as a result of the image formed on the array. After a suitable time, generally $\frac{1}{30}$ of a second, the positive potential is sequentially applied to L$_2$, then L$_3$, then back to L$_1$, while the other two lines return to ground potential. The charge is then moved sequentially to the end of the line of gates and is read.

To enhance the speed of this process, a second array, identical to the first but shielded from the incident light, is often set next to the first. At read time the charge is laterally transferred into this *transport register*. In this case readout can take place while a new image is formed under the detector array. A number of other raster array schemes are used as well, but an increase in the number of registers reduces the area available for the detector, so some balance is required.

By making the sensitivity of alternate rows red, green, and blue, one can get a color-sensitive CCD. These are now inexpensive enough that they have become the standard imaging device for home video cameras.

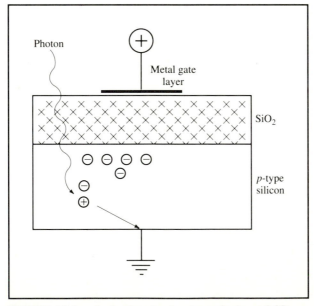

FIGURE 12.6.   A single electrode gate in a CCD.

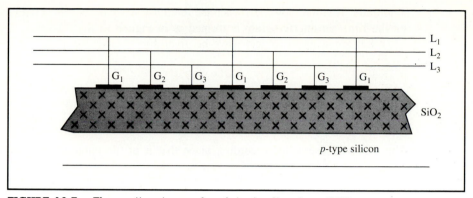

FIGURE 12.7.   The gating to read a detector line in a CCD.

The responsivity of the CCD is dependent on the wavelength of the incident light as well as its intensity and is given by R_λ, the spectral response. The input is usually a step function and R_λ depends on a time constant τ. There is also a long-wavelength cutoff at the point where the photon energy falls below that required to produce electron–hole pairs. Practically, the cutoff frequency depends on τ and is given by $R(0)/\sqrt{2}$, where $R(f)$ is

$$R(f) = \frac{R(0)}{(1 + 4\pi^2 f^2 \tau^2)^{1/2}} \tag{12.5}$$

and f is the frequency at which the light input to the device is modulated.

Holography

The concept of a holographic image was first proposed by Dennis Gabor in 1948. Its real success came only after the development of the laser, in which the highly coherent intense laser source allowed one to have two separate coherent beams from the same source and additionally provided a good source for image reconstruction. Holography is now a viable tool; in fact, many credit cards bear a holographic image in an effort to avoid counterfeiting.

The hologram differs from the photographs discussed earlier in that it contains not only amplitude information but phase information as well. The techniques of *holography,* the reconstruction of wavefronts, are sufficiently general that they may be extended throughout the electromagnetic spectrum and are even applied to acoustic waves.

To record the phase information as well as the amplitude, one needs a phase reference. This is provided by a *reference wave,* a second wave simultaneously incident on the plate with the *object wave* scattered from the object of interest. The object wave carries the information about both the object amplitude and phase. Film will record only amplitude variation; however, when the reference wave is combined with the object wave on the film, an interferogram is formed as in Figure 12.8.

Let the scattered object wave be given by

$$a(x,y) = a(x,y)e^{-j\phi(x,y)} \tag{12.6}$$

and the reference wave by

$$A(x,y) = A(x,y)e^{-j\theta(x,y)} \tag{12.7}$$

The sum of these at the recording plane that is at the photographic plate is given by[1]

$$A_{\text{fin}}(x,y) = A(x,y)e^{-j\theta(x,y)} + a(x,y)e^{-j\phi(x,y)} \tag{12.8}$$

and the recorded information, the intensity derived from this sum, is given by

$$I(x,y) = |A(x,y)|^2 + |a(x,y)|^2 + 2A(x,y)a(x,y)\cos[\theta(x,y) - \phi(x,y)] \tag{12.9}$$

The first two terms of this expression are similar to those for a normal photograph in that they contain only intensity information, while the third term contains the relative phases. This recorded information with the phase data is called a *hologram*.

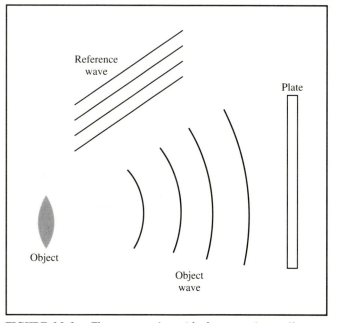

FIGURE 12.8.   The geometry of hologram formation.

---

[1] Time is not included in these expressions since both beams are coherent and assumed to be from the same source. In addition, they are recorded simultaneously.

The recording is done on a photographic plate with care being taken to ensure that the variations of the particular exposure lie in the linear portion of the H-D curve of the plate. Care must also be taken in selecting holographic plates to see that their spatial frequency response is sufficient for the particular application.

The intensity $|A|^2$ of the reference beam will be constant across the surface of the recording plate. The transmission of the recorded plate will be given by

$$t(x,y) = t_0 + \beta(|a^2| + A^*a + aA^*) \qquad (12.10)$$

where t_0 represents the constant dc bias term and β is the product of the film's Γ and the exposure time at the bias point. β is negative for a negative transparency and positive for a positive transparency.

Wavefront Reconstruction

The recorded object wave a can be recovered from the exposed and processed film by use of a *reconstructing wave*, a coherent wave $\mathbf{B}(x,y)$. The light which is transmitted by the transparency when interrogated by $\mathbf{B}(x,y)$ is given by

$$\mathbf{B}(x,y)t(x,y) = t_0\mathbf{B} + \beta aa^*\mathbf{B} + \beta A^*\mathbf{B}a + \beta AB a^* \qquad (12.11)$$

and the terms on the right-hand side of (12.11) are labeled U_1, U_2, U_3, and U_4, respectively. The third term, U_3, is of interest because, when $\mathbf{B}$ is the same as the reference wavefront $\mathbf{A}$, then U_3 becomes

$$U_3 = \beta|A^2|a \qquad (12.12)$$

and $U_3(x,y)$ is proportional to the original object wavefront as in Figure 12.9.

The fourth term, $U_4(x,y)$, under the condition that $B = A^*$, becomes

$$U_4 = \beta|A^2|a^* \qquad (12.13)$$

which is proportional to a^*, the conjugate of the object wave. This is shown in Figure 12.10.

Unfortunately it is not possible to isolate either U_3 or U_4. In addition, each is always accompanied by two other terms, U_1 and U_2, which are in essence noise terms.

Image Reconstruction

The concept of wavefront reconstruction can be extended rather directly to image formation. The reconstructed wave U_3 appears to an observer at the right of the hologram to be diverging from a position behind the hologram originally occupied by the object when the hologram was exposed as in Figure 12.11. The image is formed as in Figure 12.12 and is virtual.

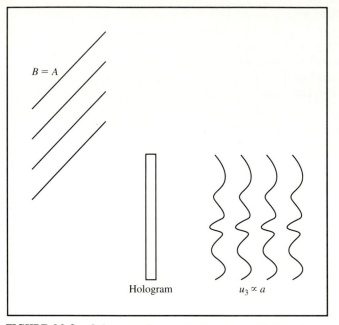

FIGURE 12.9.   Interrogating a hologram with $B = A$.

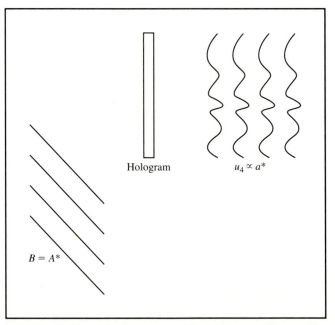

FIGURE 12.10.   Interrogating a hologram with $B = A^*$.

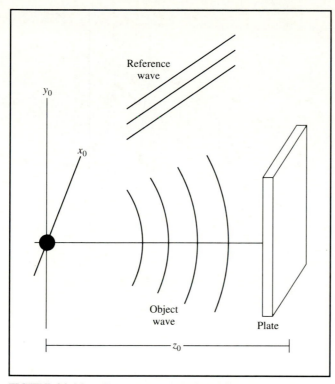

FIGURE 12.11. The formation of the hologram of a point object.

Similarly, when A^* is used to interrogate the hologram during reconstruction, U_4 also generates an image. In this case the image is real, as shown in Figure 12.13. The point object in Figure 12.11 scatters a spherical wave which can be expressed as

$$a(x,y) = a_0 e^{jk[z_0^2+(x-x_0)^2+(y-y_0)^2]^{1/2}} \tag{12.14}$$

where $(-x_0,-y_0,-z_0)$ is the position of the object. When the hologram is interrogated by A^* one gets

$$U_4 = \beta|A^2|a^*(x,y) \tag{12.15}$$

which is

$$U_4 = \beta|A^2|e^{-jk[z_0^2+(x-x_0)^2+(y-y_0)^2]^{1/2}} \tag{12.16}$$

This is a spherical wave converging to a point z_0 to the right of the hologram. An extended object can be taken as a set of point sources, so that by the linearity of the system an extended image can also be formed as a collection of point images.

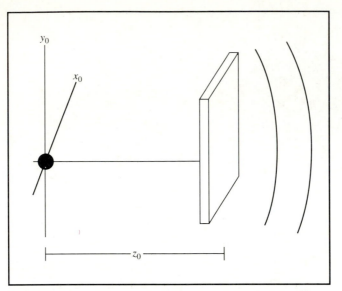

FIGURE 12.12. The virtual image formed by the illumination of the hologram with the reconstructing wave $B = A$.

The virtual image arising in U_3 is called the *orthoscopic* or *true image*, while that formed from U_4, the real image, is called the *pseudoscopic image*. The pseudoscopic image looks to the observer to be inside out. If the image were that of a face, for example, the view would appear to be from behind the head rather than from in front of it. The nose would appear to be behind the chin rather than in front of it.

When the eye forms an image, only a small portion of the wavefront scattered from the object, that portion captured by the eye's pupil, is used. Thus, in viewing a hologram, only a small portion of a wavefront is required to provide the view. This requires only a portion of the full plate. The plate can be broken apart and the pieces will still provide a reconstructed wavefront. The image that one views has all the perspective of the original scene. Objects appear to occupy their positions is space, and this gives the image a three-dimensional quality.

The emulsion in a photographic plate is of the order of 10 times thicker than 1 wavelength of light, and the effect of this thickness must be taken into account. This can be easily treated for the case where the object wave is a plane wave, as would be the case where the object is a distant point. If the normals to the object wavefront and the reference wavefront each make angles of $\phi/2$ with respect to the surface normal, one has the situation illustrated in Figure 12.14. The wavefronts, shown in the figure by the dashed lines, will interfere constructively, giving planes where the exposure is intense, as shown by the heavy vertical lines in the figure, while the regions between will be only lightly exposed. In the case illustrated one would have a grating when the plate was developed.

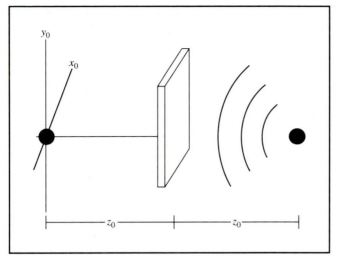

FIGURE 12.13. The formation of a real point image through the illumination of a hologram with $B = A^*$.

If the separation of the vertical maxima in Figure 12.14 is given by δ, it is easy to show that

$$2\delta \sin\frac{\phi}{2} = \lambda \qquad (12.17)$$

where λ is the wavelength of the reference beam. What angle of incidence would be used to reconstruct the object wave so that the reconstructed wave has maximum intensity? One interesting approach is to consider the vertical maxima as mirrors. Let the angle of incidence of the plane wave reconstructing beam on the "mirrors" be given by ϕ in Figure 12.15. The throughput will be in the direction shown in the figure. The maximum output will occur when the reflected waves are in phase so that from mirror to mirror the successive path length increase must be 1 wavelength. This is true only if

$$\sin \phi = \pm\frac{\lambda}{2\delta} \qquad (12.18)$$

which gives $\phi = \pm\lambda/2\delta$ by comparison with (12.17).

To get maximum brightness in the reconstruction of a hologram, the hologram should be illuminated by a replicate of the original reference beam. This, of course, is the beam which gives rise to the orthoscopic image. If the hologram is illuminated by light at $\phi = -(\pi - \theta/2)$, retrograde illumination, one gets the pseudoscopic image. The important facts are these: to get the orthoscopic image, illuminate along the same path as the reference beam, but to get the pseudoscopic image, illuminate from the opposite side of the plate. Maximally bright images will only be gotten with proper illumination.

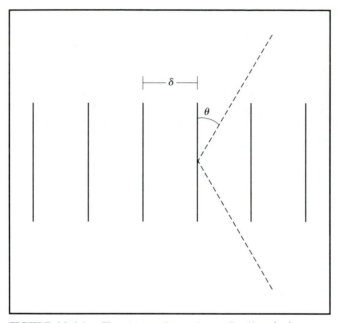

FIGURE 12.14.   The formation of a reflection hologram.

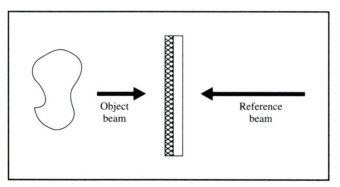

FIGURE 12.15.   A thick emulsion illuminated from the right by a reference wave and from the left by an object wave. The intense vertical lines show the plane of constructive interference in the emulsion.

White-Light Holograms

There are several kinds of holograms which can be reconstructed using white light. Among these are the *reflection hologram*, the *image hologram*, and the *rainbow hologram*. The reflection hologram differs from the holograms discussed earlier, in which the silver plates in the emulsion are arranged normal to the surface as in Figure 12.14. In white-light reflection holograms the plates are shaped so that the reflected beam forms the image. When the plates of silver in the emulsion act as reflecting mirror surfaces they are insensitive to changes in the wavelength of the interrogating illumination.

The reflection hologram is created by illuminating the plate from opposite sides by the object beam and the reference beam as in Figure 12.16. The interference between the reference beam and the object beam results in the exposed emulsion having a layered structure nearly parallel to the plate's surface. If the spacing of the planes is $\lambda/2$ and if this plate is interrogated using white light, only the wavelength satisfying Bragg's law,

$$\lambda = 2d \sin \theta \qquad\qquad (12.19)$$

will result in a reconstructed monochromatic image. The pseudoscopic image can again be gotten by illuminating the opposite side of the plate.

In practice, when the emulsion is developed it shrinks somewhat so that the spacings will tend to be smaller than one would predict based on the wavelength used in forming the hologram. This is overcome to some extent by reswelling the emulsion after fixing, using emulsion hardeners such as an alum salt or formaldehyde. With care, one can select the reflection wavelength which will give an image.

If a hologram formed as described earlier in this chapter is interrogated with light having a range of wavelengths, the angle of constructive interference of the scattered light will depend on the wavelength. The diffraction angles of the longer wavelengths will be greater than those of the shorter wavelengths, and the image will have a chromatic blur. This can be alleviated somewhat by making the image close to the film, generally by using a lens to image the illuminated object close to the film plane, giving the name image hologram.

The rainbow hologram is prepared by interrogating a hologram over only a narrow strip in such a fashion as to generate a pseudoscopic image. A second hologram is formed from this reconstructed image. Actually, the second hologram has as its object the strip of the original hologram. The newly formed second hologram is the rainbow hologram.

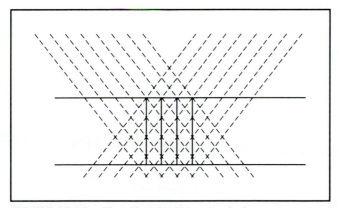

FIGURE 12.16.   The illumination of a hologram formed in a thick emulsion.

When this is interrogated with white light a series of strips of different color of the original hologram are formed. As this is viewed from different angles, the image changes color across the spectrum like a rainbow. These images are generally very bright.

Holographic Interferometry

The deformations which are visible in an object subjected to a stress are of great importance in many engineering fields. Holography provides one means of visualizing these deformations. The object of interest, which in some cases is a model of a bridge or other structure, is used as the object in the formation of a hologram in the typical experimental arrangement such as that shown in Figure 12.17. The hologram is made, but the plate remains in place while a load is applied to the object; then the plate is exposed for a second time. The plate is developed and fixed as usual and then reinserted into the plate holder on the optical table and the beam splitter is replaced by a fully silvered mirror. One can then view the fringes representing the deformation.

Speckle

One of the striking characteristics of a hologram reconstituted with coherent light is a granularity or *speckle*. This does not appear in white-light holograms and is a result of the

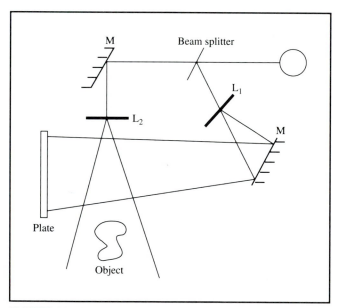

FIGURE 12.17.   The experimental arrangement for the formation of a hologram. The beam splitter has a 90:10 ratio with 10% of the output directed onto L_1. L_1 and L_2 are beam expanders. M is a front-surface mirror.

coherence of the illuminating radiation. Just as two coherent beams can interfere, the coherent radiation scattered from a surface will interfere, but the pattern will not be a series of bars; rather, it will consist of spots with intensity varying from zero to some maximum. Wherever the amplitude vector sum of the scattered radiation is zero, there is no illumination. Since intensities rather than amplitudes add in white light as the field sum is generated, this effect is not evident in white light.

Actually there are two classes of speckle: that due to rough surfaces between the source and the observer and that due to the observer's eye. The rough surface may be the surface of the illuminated object, the rough surface of a diffusing plate in the optical train, or even dust in the optical path. This roughness results in phase variation and is known as *objective speckle*. While often an annoyance, objective speckle has been used in metrology.

The second form of speckle, *subjective speckle,* arises in the eye of the viewer. The eye of an individual has a finite aperture and as a result a point-spread function. As the coherent radiation from different points on an object passes through the eye, portions of the point-spread function interfere coherently, and the retinal image then has inherent speckle. This effect has been used to evaluate visual acuities.

PROBLEMS

12.1. Why are three gates required in a CCD readout system rather than simply two? Draw a diagram of the potentials at the gates as the readout progresses.

12.2. Find an expression relating the transmittance τ to the exposure.

12.3. The amplitude transmittance for a film is given by $t = \sqrt{\tau}$. Relate this to the exposure E and to the inertia i.

12.4. Three films are rated at ASA 100, 400, and 1000. Each is given the same exposure. How will the densities compare?

12.5. Show that equation (12.9) follows from (12.8).

12.6. Discuss the origin of the sign of β in equation (12.10).

12.7. Show that equation (12.17) is valid.

12.8. Holography can provide a means for high-density data storage. Images can be stored on a hologram with high densities of information. Compare this with microfilm, for example, and suggest why it may or may not be more effective.

12.9. When the holographic interferogram formed as in Figure 12.17 is placed in the plate holder, where would one expect to find the image?

APPENDIX I

Matrices

Matrices are rectangular arrays of numbers such as

$$\begin{bmatrix} a & b \\ c & d \end{bmatrix} \qquad [7 \quad 19] \qquad \begin{bmatrix} 6 \\ 13 \end{bmatrix} \qquad \begin{bmatrix} 1 & 3 & 5 \\ 7 & 9 & 11 \end{bmatrix} \tag{AI.1}$$

The first is a 2×2 square matrix, the second a row matrix with two elements, the third a column matrix, and the fourth a rectangular matrix with two rows and three columns.

Matrices have an algebra of their own. Two matrices are equal if they have the same number of rows and columns and if each of the corresponding elements is equal. Two matrices can be added if they have the same number of rows and columns, that is, if they conform to one another. Addition takes place term by term with corresponding terms being added together. For example,

$$\begin{bmatrix} a & b \\ c & d \end{bmatrix} + \begin{bmatrix} m & n \\ r & s \end{bmatrix} = \begin{bmatrix} a+m & b+n \\ c+r & d+s \end{bmatrix} \tag{AI.2}$$

Matrices may be multiplied by a scalar number, and this results in each element of the matrix being multiplied by that scalar. For example,

$$3\begin{bmatrix} a & b \\ c & d \end{bmatrix} = \begin{bmatrix} 3a & 3b \\ 3c & 3d \end{bmatrix} \tag{AI.3}$$

Of course this also presents the possibility of factoring a common term from each element of the matrix and carrying that term as a multiplicative factor.

Two matrices may be multiplied if the number of columns of the first is equal to the number of rows of the second. Each column element of a row of the first matrix is multiplied by a corresponding row element of a column of the second, and the terms are added to give an element in the product matrix. For example, one can multiply a 2×2 matrix with two columns by a 2×1–column matrix with two rows,

$$\begin{bmatrix} a & b \\ c & d \end{bmatrix} \begin{bmatrix} r \\ s \end{bmatrix} = \begin{bmatrix} ar + bs \\ cr + ds \end{bmatrix} \tag{AI.4a}$$

to give a second column matrix, that is, one with the number of columns of the second and the number of rows of the first. The product of the two 2×2 matrices is again a 2×2 matrix.

$$\begin{bmatrix} a & b \\ c & d \end{bmatrix} \begin{bmatrix} k & l \\ m & n \end{bmatrix} = \begin{bmatrix} ak + bm & al + bn \\ ck + dm & cl + cn \end{bmatrix} \tag{AI.4b}$$

Note that reversing the order of the matrices in the first example gives a product which cannot be evaluated since then the one column of the first does not equal the two rows of the second.

Reversing the order of the matrices in the second example demonstrates the non-commutative nature of matrix multiplication. Matrix multiplication is in general not commutative, although it may be in special circumstances. That is, in general, if the products **AB** and **BA** both exist, then usually $\mathbf{AB} \neq \mathbf{BA}$.

Matrices are of great use in the solution of systems of equations. If one has

$$\begin{aligned} y_1 &= a_{11}x_1 + a_{12}x_2 \\ y_2 &= a_{21}x_1 + a_{22}x_2 \end{aligned} \tag{AI.5}$$

and

$$\begin{aligned} x_1 &= b_{11}z_1 + b_{12}z_2 \\ x_2 &= b_{21}z_1 + b_{22}z_2 \end{aligned} \tag{AI.6}$$

and if one wants to express the y values in terms of the z values in the form

$$\begin{aligned} y_1 &= c_{11}z_1 + c_{12}z_2 \\ y_2 &= c_{21}z_1 + c_{22}z_2 \end{aligned} \tag{AI.7}$$

direct substitution gives

$$\begin{aligned} y_1 &= (a_{11}b_{11} + a_{12}b_{21})z_1 + (a_{11}b_{12} + a_{12}b_{22})z_2 \\ y_2 &= (a_{21}b_{11} + a_{22}b_{21})z_1 + (a_{21}b_{12} + a_{22}b_{22})z_2 \end{aligned} \tag{AI.8}$$

If one expresses equations (AI.6), (AI.7), and (AI.8) in matrix form,

$$y = \mathbf{A}x$$
$$x = \mathbf{B}z \qquad\qquad (AI.9)$$
$$y = \mathbf{C}z$$

where $\mathbf{A}$, $\mathbf{B}$, and $\mathbf{C}$ are the matrices of the coefficients of the equations, one can see that

$$\begin{bmatrix} c_{11} & c_{12} \\ c_{21} & c_{22} \end{bmatrix} = \begin{bmatrix} a_{11} & a_{12} \\ a_{21} & a_{22} \end{bmatrix}\begin{bmatrix} b_{11} & b_{12} \\ b_{21} & b_{22} \end{bmatrix} \qquad (AI.10)$$

This is the important aspect of matrix manipulation related to ray tracing. In ray tracing one is simply substituting a series of new coordinate systems as one works one's way through the lens system.

One important matrix is the unit matrix. The 2×2 unit matrix, $\mathbf{I}$, is written

$$\begin{bmatrix} 1 & 0 \\ 0 & 1 \end{bmatrix} \qquad (AI.11)$$

This matrix commutes with any 2×2 matrix and is equivalent to multiplication by 1 so that, if $\mathbf{A}$ is a 2×2 matrix, then

$$\mathbf{AI} = \mathbf{IA} = \mathbf{A} \qquad (AI.12)$$

Determinants

Associated with each square matrix is a determinant. The determinant is a scalar value which is found for a 2×2 determinant as

$$\begin{bmatrix} a & b \\ c & d \end{bmatrix} = ad - bc \qquad (AI.13)$$

There is a useful rule in linear algebra which finds great use in paraxial ray tracing using matrices:

> *The determinant of a product of square matrices is equal to the product of the determinants of the individual matrices.*

This rule can be verified easily by direct calculation for the 2×2 case with two arbitrary matrices and can be generalized to n 2×2 matrices by mathematical induction. Let the determinants of **A** and **B** be written

$$|\mathbf{A}| = \begin{bmatrix} a_{11} & a_{12} \\ a_{21} & a_{22} \end{bmatrix} = a_{11}a_{22} - a_{12}a_{21} \tag{AI.14}$$

and

$$|\mathbf{B}| = \begin{bmatrix} b_{11} & b_{12} \\ b_{21} & b_{22} \end{bmatrix} = b_{11}b_{22} - b_{12}b_{21} \tag{AI.15}$$

The matrix product of the matrices **A** and **B** is given by

$$\mathbf{AB} = \begin{bmatrix} a_{11}b_{11} + a_{12}b_{21} & a_{11}b_{12} + a_{12}b_{22} \\ a_{21}b_{11} + a_{22}b_{21} & a_{21}b_{12} + a_{22}b_{22} \end{bmatrix} \tag{AI.16}$$

and the determinant by

$$\begin{aligned} |\mathbf{AB}| &= (a_{11}b_{11} + a_{12}b_{21})(a_{21}b_{12} + a_{22}b_{22}) - (a_{11}b_{12} + a_{12}b_{22})(a_{21}b_{11} + a_{22}b_{21}) \\ &= a_{11}a_{22}b_{11}b_{22} + a_{12}b_{21}a_{21}b_{12} - a_{11}b_{12}a_{22}b_{21} - a_{12}a_{21}b_{11}b_{22} \end{aligned} \tag{AI.17}$$

which can be factored to give

$$(a_{11}a_{22} - a_{12}a_{21})(b_{11}b_{22} - b_{12}b_{21}) \tag{AI.18}$$

which is the product of the determinants (AI.14) and (AI.15).

APPENDIX II

The Diffraction Integrals

The wave equation (6.5) was solved in Chapter 6 to show the linearity of the solution and to introduce the principles of interference arising from the superposition of optical waves. Here we will examine further consequences of the solutions of this equation.

The separated spatial part of the wave equation (6.12) can be written

$$(\nabla^2 + k^2)U(\mathbf{P}) = 0 \qquad \text{(AII.1)}$$

where $\mathbf{P}$ is a field point and $U(\mathbf{P})$ is the quantity of interest, that is, the optical disturbance at $\mathbf{P}$. This equation is known as the Helmholtz equation, and one needs its solution to find $U(\mathbf{P})$. It is a partial differential equation, and therefore its solution is dependent on the boundary conditions. One approach to the solution of this problem is to invoke Green's theorem, which is written

$$\int\int\int (G\nabla^2 U - U\nabla^2 G)\, dV = \int\int \left(G\frac{\partial U}{\partial n} - U\frac{\partial G}{\partial n} \right) dS \qquad \text{(AII.2)}$$

where $\partial/\partial n$ is the partial derivative with respect to the outward-going normal to the surface S of the volume of interest V. Careful selection of an appropriate surface S allows one to use Green's theorem to solve (AII.2).

Here the solution to be followed is that originally put forward by Kirchhoff. Figure AII.1 illustrates the geometry. The Green's function G is taken to be a unit amplitude spherical wave expanding about P_0. This is known as the free-space Green's function

$$G(P_1) = \frac{e^{jkr_{01}}}{r_{01}} \qquad \text{(AII.3)}$$

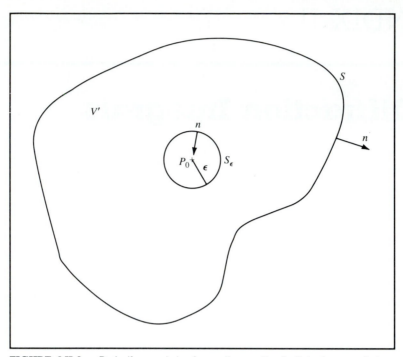

FIGURE AII.1. P_0 is the point where the optical disturbance is to
be determined, V' is the region of interest, and n represents the
outward-going normals.

where P_1 is some point at a distance r_{01} from P_0, so that in order to use the Green's
function (AII.3), one must exclude from the volume the small sphere of radius ϵ surrounding P_0 so that the surface integral (AII.2) will be over the surface $S' = S + S_\epsilon$ which
surrounds V'. Inclusion of the small volume is required since $G(P_1)$ becomes infinite at
$r_{01} = 0$. Within V' the function G satisfies Helmholtz's equation so that

$$(\nabla^2 + k^2)G = 0 \qquad\qquad (AII.4)$$

and

$$(\nabla^2 + k^2)U = 0 \qquad\qquad (AII.5)$$

Green's theorem (AII.2) gives

$$\iiint (G\nabla^2 U - U\nabla^2 G)\, dV = -\iiint (GUk^2 - UGk^2)\, dV = 0 \qquad (AII.6)$$

so that

$$\iint\limits_{S'} \left(G\frac{\partial U}{\partial n} - U\frac{\partial G}{\partial n} \right) dS = 0 \qquad\qquad (AII.7)$$

and thus

$$\iint_S \left(G\frac{\partial U}{\partial n} - U\frac{\partial G}{\partial n} \right) dS = -\iint_S \left(G\frac{\partial U}{\partial n} - U\frac{\partial G}{\partial n} \right) dS \qquad \text{(AII.8)}$$

For a point P_1 on S',

$$G(P_1) = \frac{e^{jkr_{01}}}{r_{01}} \qquad \text{(AII.9)}$$

and

$$\frac{\partial G(P_1)}{\partial n} = \cos(n, r_{01})\left(jk - \frac{1}{r_{01}} \right)\frac{e^{jkr_{01}}}{r_{01}} \qquad \text{(AII.10)}$$

where $\cos(n, r_{01})$ is the cosine of the angle between n and r_{01}. On S_ϵ one has $\cos(n, r_{01}) = -1$ and

$$G(P_1) = \frac{e^{jk\epsilon}}{\epsilon} \qquad \text{(AII.11)}$$

$$\frac{\partial G(P_1)}{\partial n} = \frac{e^{jk\epsilon}}{\epsilon}\left(\frac{1}{\epsilon} - jk \right) \qquad \text{(AII.12)}$$

One assumes that U and its derivatives are continuous at P_0 and lets $\epsilon \to 0$ so that

$$\iint_{S_\epsilon} \left(G\frac{\partial U}{\partial n} - U\frac{\partial G}{\partial n} \right) dS = 4\pi\epsilon^2 \left[\frac{\partial(P_0)}{\partial n}\frac{e^{jk\epsilon}}{\epsilon} - U(P_0)\frac{e^{jk\epsilon}}{\epsilon}\left(\frac{1}{\epsilon} - jk \right) \right]$$

$$= -4\pi U(P_0) \qquad \text{(AII.13)}$$

as $\epsilon \to 0$. One then has

$$U(P_0) = \frac{1}{4\pi}\iint_S \left(\frac{\partial U}{\partial n}\frac{e^{jkr_{01}}}{r_{01}} - U\frac{\partial}{\partial n}\frac{e^{jkr_{01}}}{r_{01}} \right) dS \qquad \text{(AII.14)}$$

giving U at P_0 in terms of its value on the boundary. This is the *Kirchhoff–Helmholtz integral theorem*. The solution of the Helmholtz equation in this case then reduces to the solution of an integral over a region surrounding the point of interest.

The Fresnel–Kirchhoff Diffraction Expression

The application of the Kirchhoff–Helmholtz integral requires that we specify the surface S over which the integral is taken. Figure AII.2 represents one possible surface configuration where there is an aperture, Σ, in an opaque screen which is illuminated from the left.

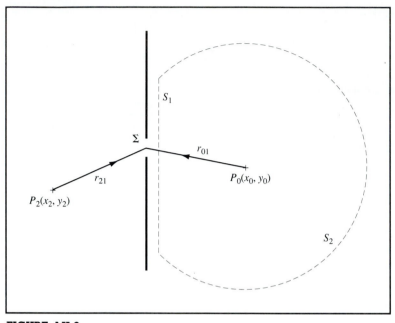

FIGURE AII.2.

The surface S is taken in two parts: the plane behind the screen designated S_1 and the spherical segment S_2 about P_0. Equation (AII.14) gives

$$U(P_0) = \frac{1}{4\pi} \iint_{S_1+S_2} \left(G\frac{\partial U}{\partial n} - U\frac{\partial G}{\partial n} \right) dS \qquad \text{(AII.15)}$$

We will again take the free-space Green's function G. As R increases, S_2 becomes a large hemisphere whose area is proportional to R^2. On R

$$G = \frac{e^{jkR}}{R} \qquad \text{(AII.16)}$$

and

$$\frac{\partial G}{\partial n} = \left(jk - \frac{1}{R} \right)\frac{e^{jkR}}{R} \approx jkG \qquad \text{(AII.17)}$$

for R very large. From (AII.15) one has

$$\iint_{S_2} \left(G\frac{\partial U}{\partial n} - jkGU \right) dS = \int_{\Omega} G\left(\frac{\partial U}{\partial n} - jkU \right)R^2\, d\omega \qquad \text{(AII.18)}$$

where Ω is the solid angle subtended by S_2 at P_0. One makes the assumption that $|RG|$ is uniformly bounded on S_2 so that

$$\lim_{R \to \infty} R\left(\frac{\partial U}{\partial n} - jkU\right) = 0 \tag{AII.19}$$

uniformly with angle. This is the *Sommerfeld radiation condition*.

For the plane S_1 the contribution to the disturbance U at P_0 comes essentially from Σ, and one takes U and $\partial U/\partial n$ on Σ to be the same as in the absence of the screen.[1] The other regions of S_1, that is, those parts of S_1 behind the screen and not in the region Σ, have both U and $\partial U/\partial n$ identically zero. This is the *Kirchhoff boundary condition*. One then has

$$U(P_0) = \frac{1}{4\pi}\iint_{\Sigma} \left(\frac{\partial U}{\partial n}G - U\frac{\partial G}{\partial n}\right) dS \tag{AII.20}$$

The point P_0 is taken many wavelengths distant from Σ, so that $k \gg 1/r_{01}$ and

$$\frac{\partial G(P_1)}{\partial n} = \cos(n,r_{01})\left(jk - \frac{1}{r_{01}}\right)\frac{e^{jkr_{01}}}{r_{01}} \tag{AII.21}$$

It remains to define U on the region Σ. For this we take a point source at P_2 so that Σ is illuminated by a spherical wave

$$jk \, \cos(n,r_{01})\frac{e^{jkr_{01}}}{r_{01}} \tag{AII.22}$$

$$U(P_0) = \frac{1}{4\pi}\iint_{\Sigma} \frac{e^{jkr_{01}}}{r_{01}}\left[\frac{\partial U}{\partial n} - jkU \, \cos(n,r_{01})\right] dS \tag{AII.23}$$

$$U(P_1) = A\frac{e^{jkr_{21}}}{r_{21}} \tag{AII.24}$$

and the quantity of interest $U(P_0)$ is then

$$U(P_0) = \frac{A}{j\lambda}\iint \left[\frac{e^{jk(r_{21}+r_{01})}}{r_{21}r_{01}}\frac{\cos(n,r_{01}) - \cos(n,r_{21})}{2}\right] dS \tag{AII.25}$$

---

[1] Strictly speaking, one must take into account the effect of the edges of the aperture on the field within the aperture. This is a difficult process at best, and for our purposes we can neglect any effect due to the fringing fields at the edge of the aperture.

where the integral is understood to be zero except over Σ. This is the *Kirchhoff–Fresnel equation*. The field at P_0 can then be thought of as arising from secondary sources at P_1 in the sense

$$U(P_0) = \iint_\Sigma U'(P_1) \frac{e^{jkr_{01}}}{r_{01}} \, dS \qquad (\text{AII.26})$$

where

$$U'(P_1) = \frac{A}{2j\lambda} \frac{e^{jkr_{21}}}{r_{21}} [\cos(n,r_{01}) - \cos(n,r_{21})] \qquad (\text{AII.27})$$

This differs from the field at P_1 due to the original source in that its amplitude has a $1/\lambda$ weighting, there is an obliquity factor with the cosine terms, and there is a phase lead of $90°$ arising in the $1/j$ term.

The evaluation of the optical disturbance at P_0 is quite difficult if one uses (AII.25). There are, however, some approximations which may be made and which reduce the difficulty in solving for the optical disturbance at P_0. Equation (AII.25) can be written

$$U(x_0,y_0) = \iint_\Sigma h(x_0,y_0; x_1,y_1) U(x_1,y_1) \, dx_1 \, dy_1 \qquad (\text{AII.28})$$

where $h(x_0,y_0; x_1,y_1)$ is given by

$$h(x_0,y_0; x_1,y_1) = \frac{1}{j\lambda} \frac{e^{jkr_{01}}}{r_{01}} \cos(n,r_{01}) \qquad (\text{AII.29})$$

Note that (AII.28) is often written with infinite limits on the integral even though U vanishes except on Σ.

The following approximations are usually made:

1. The distance r_{01} from Σ is much greater than the dimensions of Σ.

2. The region of interest is restricted to a finite region about z so that $\cos(n,r_{01})$ can be taken as 1. Note that the error arising from this assumption is $<5\%$ if the angle is restricted to a region $<18°$ about the z axis.

3. The quantity r_{01} in the denominator is taken as z so that

$$h(x_0,y_0; x_1,y_1) = \frac{1}{j\lambda z} e^{jkr_{01}} \qquad (\text{AII.30})$$

The quantity r_{01} in the exponential cannot be replaced by z since k is large, so that even small differences between r_{01} and z can lead to substantive error.

The quantity r_{01} can be expanded by taking

$$r_{01} = \sqrt{z^2 + (x_0 - x_1)^2 + (y_0 - y_1)^2}$$

$$= z\left[1 + \left(\frac{x_0 - x_1}{z}\right)^2 + \left(\frac{y_0 - y_1}{z}\right)^2\right]^{1/2} \tag{AII.31}$$

This expression can be expanded by the binomial theorem[2] and only the first two terms retained,

$$r_{01} = z\left[1 + \frac{1}{2}\left(\frac{(x_0 - x_1)}{z}\right)^2 + \frac{1}{2}\left(\frac{(y_0 - y_1)}{z}\right)^2\right] \tag{AII.32}$$

and this gives the *Fresnel approximation*

$$h(x_0, y_0; x_1, y_1) = \frac{e^{jkz}}{j\lambda z}e^{j(k/2z^2)((x_0 - x_1)^2 + (y_0 - y_1)^2)} \tag{AII.33}$$

The Fresnel approximation is valid whenever the foregoing assumptions are valid. This should be clear except perhaps for the binomial approximation assumption, for that one requires that the first term to be truncated be much less than 1 radian, which gives

$$z^3 \gg \frac{\pi}{4\lambda}[(x_0 - x_1)^2 + (y_0 - y_1)^2]_{\text{max}}^2 \tag{AII.34}$$

In this approximation,

$$U(x_0, y_0) = \frac{Ae^{2jkz}}{j\lambda z}e^{j(k/2z)(x_0^2 + y_0^2)}\int\int_{-\infty}^{+\infty} U(x_1, y_1)e^{j(k/2z)(x_1^2 + y_1^2)}e^{-j(2\pi/\lambda z)(x_0 x_1 + y_0 y_1)} \, dx_1 \, dy_1 \tag{AII.35}$$

This is the *Fresnel region diffraction equation*. One should note that $U(x_0, y_0)$ can be found from the two-dimensional Fourier transform[3] of the quantity

$$K = U(x_1, y_1)e^{j(k/2z)(x_1^2 + y_1^2)} \tag{AII.36}$$

where the frequency terms are *spatial frequencies* given by

$$f_x = \frac{x_0}{\lambda z} \qquad f_y = \frac{y_0}{\lambda z} \tag{AII.37}$$

---

[2] The binomial theorem is given by $(1 \pm b)^{1/2} = 1 \pm \frac{1}{2}b + \frac{1}{8}b^2 \pm \cdots$.

[3] See Appendix III.

and have the units reciprocal length; this can be seen from

$$U(x_0,y_0) = \frac{Ae^{jkz}}{j\lambda z}e^{j(k/2z)(x_0^2+y_0^2)}\int_{-\infty}^{+\infty}\!\!\!\int Ke^{-2\pi j(x_1 f_x + y_1 f_y)}\,dx_1\,dy_1 \qquad \text{(AII.38)}$$

The kernel K is just a "phase-weighted" distribution of the optical disturbance in Σ.

The essential aspect of the Fresnel diffraction theory is the phase weighting of the disturbance, so that in treating diffraction phenomena, particularly those relatively close to Σ, the phase is significant. All Fresnel diffraction treatments such as the Cornu spiral of Fresnel zones are based on these phase weightings.

Consider now a plane wave falling on the aperture. In this case $U(x_1,y_1)$ will be unity and (AII.35) will be separable as

$$U(x_0,y_0) = \frac{Ae^{jkz}}{j\lambda z}\int_{-\infty}^{+\infty} e^{j(\pi/2\lambda z)(x_0 - x_1)^2}\,dx_1\int_{-\infty}^{+\infty} e^{j(\pi/2\lambda z)(y_0 - y_1)^2}\,dy_1 \qquad \text{(AII.39)}$$

The forms of the integrals are identical. With the change in variables

$$x' = \frac{x_0 - x_1}{\sqrt{\lambda z}} \qquad \text{(AII.40)}$$

one can write

$$\int e^{j(\pi/2)(x')^2}\,dx' = \int_{x_L}^{x_U}\cos\frac{\pi}{2}(x')^2\,dx' + j\int_{x_L}^{x_U}\sin\frac{\pi}{2}(x')^2\,dx' \qquad \text{(AII.41)}$$

The two integrals on the right are the *Fresnel integrals*. They are not simple expressions and their values are generally given in tabular form. Table 8.1 is a short table of values of these integrals. While the limits of integration are usually taken as infinity, the integrals themselves have zero value outside the aperture, and in (AII.41) the lower and upper limits of the aperture are written as x_L and x_U, respectively. Calling

$$\int_{u_L}^{u_U}\cos\frac{\pi}{2}(u')^2\,du' = C(u') \qquad \int_{u_L}^{u_U}\sin\frac{\pi}{2}(u')^2\,du' = S(u') \qquad \text{(AII.42)}$$

equation (AII.39) becomes

$$U(x_0,y_0) = \frac{A}{2}je^{jkz}\left(\begin{array}{c}\{[C(x'_U) - C(x'_L)] - j[S(x'_U) - S(x'_L)]\}\\ \times\{[C(y'_U) - C(y'_L)] - j[S(x'_U) - S(x'_L)]\}\end{array}\right) \qquad \text{(AII.43)}$$

The corresponding intensity at (x_0,y_0,z), which is the quantity one observes, is given by

$$I(x_0,y_0,z) = \frac{A^2}{4}\left(\begin{array}{c}\{[C(x_U) - C(x_L)]^2 + [S(x_U) - S(x_L)]^2\}\\ \times\{[C(y_U) - C(y_L)]^2 + [S(y_U) - S(y_L)]^2\}\end{array}\right) \qquad \text{(AII.44)}$$

If one now returns to equation (AII.29) giving $h(x_0,y_0; x_1,y_1)$ and operates in a region where

$$z \gg \frac{k(x_1^2 + y_1^2)_{max}}{2} \tag{AII.45}$$

$U(x_0,y_0)$ can be written

$$U(x_0,y_0) = \frac{e^{jkz}}{j\lambda z} e^{j(k/2z)(x_0^2 + y_0^2)} \int\!\!\int_{-\infty}^{+\infty} U(x_1,y_1) e^{-2\pi j(f_x x_1 + f_y y_1)} \, dx_1 \, dy_1 \tag{AII.46}$$

and the phase weighting in Σ is now removed. Clearly for this to occur one must be quite far from Σ. If the slit has dimensions of 1 mm, then $z \gg 10^6$ m, a much more stringent requirement than for Fresnel diffraction. This is the Fraunhofer approximation and equation (AII.46) is the *Fraunhofer diffraction equation*. Except for some amplitude weightings in the coefficient of the integral, the Fraunhofer diffraction pattern is given by the Fourier transform of the disturbance in the aperture.

APPENDIX III

The Fourier Transform

One computational technique which finds extensive application throughout science and engineering is the Fourier transform. This technique is derived from Fourier's studies of heat transfer, but it can be applied wherever one wants to express an aperiodic function as an expansion over a continuous range of frequencies. This appendix is not intended to be either rigorous or complete; rather, it will just present the theorem along with several important results. A short table of transforms is included.

The Fourier transform of a function $f(x)$ can be represented by:

$$A(\omega) = \frac{1}{\sqrt{2\pi}} \int_{-\infty}^{+\infty} f(x) e^{-j\omega x} \, dx \qquad \text{(AIII.1a)}$$

$$B(\omega) = \int_{-\infty}^{+\infty} f(x) e^{-j\omega x} \, dx \qquad \text{(AIII.1b)}$$

$$F(\nu) = \int_{-\infty}^{+\infty} f(x) e^{-j2\pi\nu x} \, dx \qquad \text{(AIII.1c)}$$

and the inverse transforms by

$$f(x) = \frac{1}{\sqrt{2\pi}} \int_{-\infty}^{+\infty} A(\omega) e^{j\omega x} \, d\omega \qquad \text{(AIII.2a)}$$

$$f(x) = \frac{1}{2\pi} \int_{-\infty}^{+\infty} B(\omega) e^{j\omega x} \, d\omega \qquad \text{(AIII.2b)}$$

$$f(x) = \int_{-\infty}^{+\infty} F(\nu) e^{j2\pi\nu x} \, d\nu \qquad \text{(AIII.2c)}$$

Equations (AIII.1) and their corresponding inverse transforms (AIII.2) represent three of the possible ways in which the Fourier transform may be defined. Throughout this book the pair labeled (AIII.1c) and (AIII.2c) are used. This pair has the advantage that one never needs to place a coefficient on either the direct or inverse integral. Using this pair of terms, Fourier's theorem is written

$$f(x) = \int_{-\infty}^{+\infty} e^{j2\pi\nu x} \left[\int_{-\infty}^{+\infty} f(x)e^{-j2\pi\nu x}\,dx \right] d\nu \qquad (\text{AIII.3})$$

For simple functions it is usually a straightforward process to generate the transform. Consider, for example, the square pulse shown in Figure AIII.1. The transform is given by

$$F(\nu) = \mathcal{F}[f(x)] = A \int_{-\tau/2}^{+\tau/2} e^{-j2\pi\nu x}\,dx$$

and

$$F(\nu) = \left[\frac{A}{-j2\pi\nu} \right] [e^{-j\pi\nu\tau} - e^{j2\pi\nu\tau}] = A\tau \frac{\sin \pi\nu\tau}{\pi\nu\tau}$$

which is usually written $\text{sinc}(\nu\tau)$.

Table AIII.1 lists the transforms of a number of important functions.

Two-dimensional functions can often be represented as the product of two one-dimensional functions

$$f(x,y) = f(x)f(y) \qquad (\text{AIII.4})$$

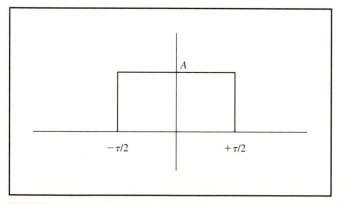

FIGURE AIII.1.   The square pulse.

TABLE AIII.1 Fourier Transform Pairs

$f(x)$	$\mathscr{F}[f(x)]$
1	$\delta(\nu)$
$\delta(x)$	1
$\delta(x \pm x_0)$	$e^{\pm j2\pi x\nu}$
$e^{\pm j2\pi x\nu}$	$\delta(\nu - \nu_0)$
$\sin(2\pi\nu_0 x)$	$(\frac{1}{2}j)[\delta(\nu - \nu_0) + \delta(\nu + \nu_0)]$
$\cos(2\pi\nu_0 x)$	$(\frac{1}{2})[\delta(\nu - \nu_0) + \delta(\nu + \nu_0)]$
$\mathrm{rect}(x) = \begin{cases} A, & \|x\| \le \dfrac{\tau}{2} \\[2mm] 0, & \|x\| > \dfrac{\tau}{2} \end{cases}$	$A\tau\,\mathrm{sinc}(\nu\tau)$
$e^{\pm j\pi x^2}$	$e^{\pm j\pi/4} e^{\mp j\pi\nu^2}$
$\mathrm{comb}(x) = \displaystyle\sum_{n=-\infty}^{n=+\infty} \delta(x - n)$	$\mathrm{comb}(v)$

In this case the transform $\mathscr{F}[f(x,y)] = \mathscr{F}[f(x)]\mathscr{F}[f(y)]$. There is one important case which appears frequently in optics where this procedure cannot be used, namely, in the case of the cylinder function $\mathrm{cyl}(r/d)$:

$$\mathrm{cyl}\left(\frac{r}{d}\right) = \begin{cases} 1, & 0 \le r < \dfrac{d}{2} \\[3mm] \dfrac{1}{2}, & r = \dfrac{d}{2} \\[3mm] 0, & r > \dfrac{d}{2} \end{cases} \qquad\text{(AIII.5)}$$

Functions such as this with circular symmetry occur frequently in optics problems. Such functions as

$$f(x,y) = g(\sqrt{x^2 + y^2}) \qquad\text{(AIII.6)}$$

clearly reflect this symmetry. The Fourier transform

$$f(\nu,\eta) = \int\!\!\!\int_{-\infty}^{+\infty} f(x,y)e^{-j2\pi(x\nu+y\eta)}\,dx\,dy \qquad\text{(AIII.7)}$$

can be written using (AIII.6) as

$$F(\nu,\eta) = \int\!\!\int_{-\infty}^{+\infty} g(\sqrt{x^2 + y^2})e^{-j2\pi(x\nu+y\eta)}\,dx\,dy \qquad \text{(AIII.8)}$$

This function can be treated in cylindrical coordinates by substituting

$$
\begin{array}{ll}
x = r\cos\theta & y = r\sin\theta \\
r = \sqrt{x^2 + y^2} & \theta = \tan^{-1}(y/x) \\
\nu = \rho\cos\phi & \eta = \rho\sin\phi \\
\rho = \sqrt{\nu^2 + \eta^2} & \phi = \tan^{-1}(\eta/\nu)
\end{array}
\qquad \text{(AIII.9)}
$$

The Fourier transform (AIII.8) with these substitutions is given by

$$F(\nu,\eta) = \int_0^\infty \int_0^{2\pi} g(r)e^{-j2\pi\rho r\cos(\theta-\phi)}r\,dr\,d\theta \qquad \text{(AIII.10)}$$

The integral on θ can be reduced using the following Bessel function identity:

$$\int_0^{2\pi} e^{-jk\cos(\theta-\phi)}\,d\theta = 2\pi J_0(k) \qquad \text{(AIII.11)}$$

where $J_0(k)$ is the zeroth-order Bessel function of the first kind.[1] When this is put into (AIII.10), one has

$$F(\nu,\eta) = 2\pi \int_0^\infty g(r)J_0(2\pi r\rho)r\,dr = F(\rho) \qquad \text{(AIII.12)}$$

which is a function only of ρ. This expansion is called a *zero-order Hankel transform* and is the cylindrical analog of the Fourier transform. The inverse transform is given by

$$g(r) = 2\pi \int_0^\infty F(\rho)J_0(2\pi r\rho)\rho\,d\rho \qquad \text{(AIII.13)}$$

The zero-order Hankel transform of (AIII.5) is given by a function called the *sombrero function*

$$\mathcal{F}[\text{cyl}(r)] = \frac{\pi}{4}\,\text{somb}(\rho) = \frac{\pi}{2}\frac{J_1(\pi\rho)}{\pi\rho} \qquad \text{(AIII.14)}$$

---

[1] See P. M. Morse and H. Feshback, *Methods of Theoretical Physics* (New York: McGraw-Hill, 1953), Vol. I, p. 621.

where J_1 is the first-order Bessel function. The sombrero function is similar to the sinc function but has different zeros, as shown in Figure AIII.2. A second zeroth-order Hankel transform of importance is given by

$$\mathcal{F}\left[\frac{\delta(r)}{\pi r}\right] = 1 \qquad \text{(AIII.15)}$$

similar to the result for the one-dimensional transform of the delta function given in Table AIII.1.

There are a number of important properties of the Fourier transform which can be used to find the Fourier transforms of a large number of functions beyond those in Table AIII.1.

Linearity:

$$\mathcal{F}[A_1 f_1(x) + A_2 f_2(x)] = A_1 \mathcal{F}[f_1(x)] + \mathcal{F}[f_2(x)] \qquad \text{(AIII.16)}$$

Scaling:

$$\mathcal{F}\left[f\left(\frac{x}{a}\right)\right] = |a| F(a\nu) \qquad \text{(AIII.17)}$$

Shift:

$$\mathcal{F}[f(x - x_0)] = e^{j2\pi x_0 \nu} F(\nu) \qquad \text{(AIII.18)}$$

Central integral:

$$F(0) = \int_{-\infty}^{+\infty} f(x) \, dx \qquad f(0) = \int_{-\infty}^{+\infty} F(\nu) \, d\nu \qquad \text{(AIII.19)}$$

FIGURE AIII.2.

Differentiation:

$$\mathcal{F}\left[\frac{d^n f(x)}{dx^n}\right] = (j\nu)^n F(\nu) \qquad\qquad \text{(AIII.20)}$$

Convolution

The *convolution* of two functions $f(x)$ and $g(x)$ is defined as

$$f(x) * g(x) = \int_{-\infty}^{+\infty} f(\alpha) g(x - \alpha) \, d\alpha \qquad\qquad \text{(AIII.21)}$$

and finds wide application in the theory of linear systems. The Fourier transform of this function,

$$\mathcal{F}[f(x) * g(x)] = \int_{-\infty}^{+\infty} e^{-j2\pi\nu x}\left[\int_{-\infty}^{+\infty} f(\alpha) g(x - \alpha) \, d\alpha\right] dx \qquad\qquad \text{(AIII.22)}$$

with the substitution $x - \alpha = \beta$ can be rewritten as

$$\int\!\!\int_{-\infty}^{+\infty} e^{-j2\pi\nu\alpha} e^{-j2\pi\nu\beta} f(\alpha) f(\beta) \, d\alpha \, d\beta = \mathcal{F}[f(x)]\mathcal{F}[g(x)] \qquad\qquad \text{(AIII.23)}$$

so that

$$\mathcal{F}[f(x) * g(x)] = F(\nu)G(\nu) \qquad\qquad \text{(AIII.24)}$$

This result is known as the *convolution or product theorem.* Similarly,

$$\mathcal{F}[f(x)g(x)] = F(\nu) * G(\nu) \qquad\qquad \text{(AIII.25)}$$

and this proves useful often in finding transforms of more complex functions.

Delta Functions

One important quantity which appears frequently in optics as well as in other linear systems is the *unit impulse function* or *delta function*. This function is used for example to model a point source of light. It has the properties

$$\delta(x - x_0) = \begin{cases} 0, & x \neq x_0 \\ 1, & x = x_0 \end{cases} \qquad \text{(AIII.26)}$$

and

$$\int_{-\infty}^{+\infty} \delta(x - x_0)f(x)\ dx = f(x_0) \qquad \text{(AIII.27)}$$

while

$$\int_{-\infty}^{+\infty} \delta(x - x_0)\ dx = 1 \qquad \text{(AIII.28)}$$

The delta function has the scaling properties

$$\delta\left(\frac{x - x_0}{a}\right) = |a|\delta(x - x_0)$$

$$\delta(ax - x_0) = \frac{1}{|a|}\delta\left(x - \frac{x_0}{a}\right) \qquad \text{(AIII.29)}$$

In two dimensions the delta function in cartesian coordinates can be written as a product

$$\delta(x - x_0, y - y_0) = \delta(x - x_0)\delta(y - y_0) \qquad \text{(AIII.30)}$$

while in polar coordinates

$$\delta(r - r_0) = \frac{\delta(r - r_0)}{r_0}\delta(\theta - \theta_0) \qquad \text{(AIII.31)}$$

For a delta function at the origin,

$$\delta(r - 0) = \frac{\delta(r)}{r\pi} \qquad \text{(AIII.32)}$$

APPENDIX IV

Waves in Material Media

It was shown in Chapter 6 that Maxwell's equations give rise to a wave equation and that the solutions of that wave equation are harmonic functions of the form

$$\mathbf{E}(\mathbf{r},t) = \mathbf{E}_s(\mathbf{r})e^{j\omega t} \tag{AIV.1}$$

and

$$\mathbf{H}(\mathbf{r},t) = \mathbf{H}_s(\mathbf{r})e^{j\omega t} \tag{AIV.2}$$

where the electric and magnetic fields $\mathbf{E}_s(\mathbf{r})$ and $\mathbf{H}_s(\mathbf{r})$ depend only on the space variables and not on time. If this solution is put into Maxwell's equations (6.1) in a charge-free region,

$$\nabla \cdot \mathbf{E}_s = 0 \tag{AIV.3a}$$

$$\nabla \cdot \mathbf{H}_s = 0 \tag{AIV.3b}$$

$$\nabla \times \mathbf{E}_s = -j\omega\mu\mathbf{H}_s \tag{AIV.3c}$$

$$\nabla \times \mathbf{H}_s = (\sigma + j\omega\epsilon)\mathbf{E}_s \tag{AIV.3d}$$

where μ is the permeability of the region, ϵ is the permittivity of the region, and σ is the conductivity defined by

$$\mathbf{J} = \sigma\mathbf{E} \tag{AIV.4}$$

Taking the curl of both sides of equation (AIV.3c) and using the vector identity

$$\nabla \times \nabla \times \mathbf{A} = \nabla(\nabla \cdot \mathbf{A}) - \nabla^2\mathbf{A} \tag{AIV.5}$$

one gets

$$\nabla^2 \mathbf{E}_s - j\mu\omega(\sigma + j\omega\epsilon)\mathbf{E}_s = 0 \qquad (AIV.6)$$

which is usually written in the form

$$\nabla^2 \mathbf{E}_s - \gamma^2 \mathbf{E}_s = 0 \qquad (AIV.7)$$

where γ is the *propagation constant* of the medium, a complex quantity of the form

$$\gamma = \alpha + j\beta \qquad (AIV.8)$$

where

$$\gamma^2 = j\mu\omega(\sigma + j\omega\epsilon) \qquad (AIV.9)$$

A little algebra shows that

$$\alpha = \omega\sqrt{\frac{\mu\epsilon}{2}\left[\sqrt{1 + \left(\frac{\sigma}{\omega\epsilon}\right)^2} - 1\right]} \qquad (AIV.10)$$

and

$$\beta = \omega\sqrt{\frac{\mu\epsilon}{2}\left[\sqrt{1 + \left(\frac{\sigma}{\omega\epsilon}\right)^2} + 1\right]} \qquad (AIV.11)$$

If one has a wave polarized in the x direction and propagating in the z direction, the solution takes the form

$$\mathbf{E}(z,t) = E_0 e^{-\alpha z} e^{j(\omega t - \beta z)} \hat{a}_x \qquad (AIV.12)$$

where $\hat{a}_x$ is a unit vector in the x direction illustrating the fact that the amplitude is damped in this medium. Substituting this result in equation (AIV.3) shows that

$$\mathbf{H}(z,t) = H_0 e^{-\alpha z} e^{j(\omega t - \beta z)} \hat{a}_y \qquad (AIV.13)$$

and

$$H_0 = \frac{E_0}{\eta} \qquad (AIV.14)$$

where

$$\eta = \sqrt{\frac{j\omega\mu}{\sigma + j\omega\epsilon}} \qquad (AIV.15)$$

The quantity η, known as the intrinsic impedance, can be written in the form

$$\eta = |\eta|e^{j\theta_\eta} \tag{AIV.16}$$

where

$$|\eta| = \frac{\sqrt{\mu/\epsilon}}{\sqrt[4]{1 + [\sigma/\omega\epsilon]^2}} \tag{AIV.17}$$

and

$$\tan 2\theta_\eta = \frac{\sigma}{\omega\epsilon} \tag{AIV.18}$$

If one compares equation (AIV.12) with equation (6.13), one immediately sees that β is the wave number and the velocity of propagation is then given by

$$v = \frac{\omega}{\beta} \tag{AIV.19}$$

In free space $\sigma = 0$ and η becomes the purely real quantity $\sqrt{(\mu_0/\epsilon_0)} \approx 377 \ \Omega$. Also in free space

$$\beta = \omega\sqrt{\mu_0\epsilon_0} = \frac{\omega}{c} \quad \text{so that} \quad v = \frac{1}{\sqrt{\mu_0\epsilon_0}} = c \tag{AIV.20}$$

In optics the materials of interest are those such as glass, where the velocity of propagation is altered but where the attenuation within the medium is quite small. Such materials are called *loss-less media* and are characterized by

$$\sigma \simeq 0 \quad \epsilon = \epsilon_0\epsilon_r \quad \mu = \mu_0\mu_r \tag{AIV.21}$$

where ϵ_r and μ_r are the relative permittivity and permeability, respectively. In this case $\eta = \sqrt{(\mu/\epsilon)}$ and $\mathbf{E}$ and $\mathbf{H}$ are in phase.

Reflection

One question of significance is that of a change of medium when a wave moving in one homogeneous, isotropic medium strikes an interface separating that medium from a second homogeneous, isotropic medium as in Figure AIV.1. For convenience the interface is taken as the $z = 0$ plane and the wave will travel in the z direction. In medium 1 the $\mathbf{E}$

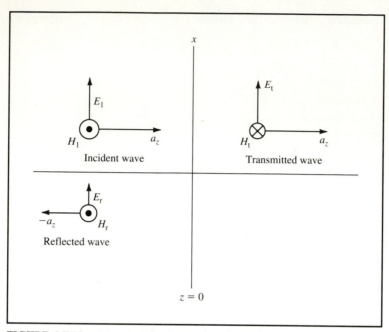

FIGURE AIV.1. The incident, reflected, and transmitted components for a wave normally incident on an interface between two homogeneous, isotropic media.

wave will be taken as polarized in the x direction so that most generally one has, with the time term omitted,

$$\mathbf{E}_{is} = E_{i0}e^{-\gamma_1 z}\hat{a}_x \qquad (\text{AIV.22})$$

and

$$\mathbf{H}_{is} = \frac{E_{i0}}{\eta_1}e^{-\gamma_1 z}\hat{a}_y \qquad (\text{AIV.23})$$

The reflected wave will travel in the negative z direction in medium 1, and

$$\mathbf{E}_{rs}(z) = E_{r0}e^{+\gamma_1 z}\hat{a}_x \qquad \mathbf{H}_{rs}(z) = -\frac{E_{r0}}{\eta_1}e^{+\gamma_1 z}\hat{a}_y \qquad (\text{AIV.24})$$

There will be a transmitted wave in medium 2 and

$$\mathbf{E}_{ts} = E_{t0}e^{-\gamma_2 z}\hat{a}_x \qquad \mathbf{H}_{ts} = \frac{E_{t0}}{\eta_2}e^{-\gamma_2 z}\hat{a}_y \qquad (\text{AIV.25})$$

and in medium 2 this is the only wave. The fields in medium 1 are given by

$$\mathbf{E}_1 = \mathbf{E}_i + \mathbf{E}_r \qquad \mathbf{H}_1 = \mathbf{H}_i + \mathbf{H}_r \tag{AIV.26}$$

while in medium 2, $\mathbf{E}_2 = \mathbf{E}_t$ and $\mathbf{H}_2 = \mathbf{H}_t$. The boundary conditions for an electromagnetic field require that the tangential components of $\mathbf{E}$ and $\mathbf{H}$ must be continuous. Since the wave is transverse and the direction of propagation is normal to the surface, the boundary conditions give

$$\mathbf{E}_i(0) + \mathbf{E}_r(0) = \mathbf{E}_t(0) \qquad \mathbf{H}_i(0) + \mathbf{H}_r(0) = \mathbf{H}_t(0) \tag{AIV.27}$$

or

$$E_{i0} + E_{r0} = E_{t0} \qquad H_{i0} + H_{r0} = H_{t0}$$

so that

$$\frac{1}{\eta_1}(E_{i0} - E_{r0}) = \frac{E_{t0}}{\eta_2} \tag{AIV.28}$$

and

$$E_{r0} = \frac{\eta_2 - \eta_1}{\eta_2 + \eta_1} E_{i0} \tag{AIV.29}$$

$$E_{t0} = \frac{2\eta_2}{\eta_2 + \eta_1} E_{i0} \tag{AIV.30}$$

The *reflection coefficient* Γ is defined as

$$\Gamma = \frac{E_{r0}}{E_{i0}} = \frac{\eta_2 - \eta_1}{\eta_2 + \eta_1} \tag{AIV.31}$$

and the *transmission coefficient* τ as

$$\tau = \frac{E_{t0}}{E_{i0}} = \frac{2\eta_2}{\eta_2 + \eta_1} \tag{AIV.32}$$

where $1 + \Gamma = \tau$ with both Γ and τ dimensionless and $0 \leq \Gamma \leq 1$.

One can see that if medium 2 is a conductor so that $\eta_2 = 0$, there is no transmitted wave, only a reflected wave. Metals are therefore good mirrors!

What is detected is not the amplitude of the field but its intensity EE^*. The *reflectivity*, R, is given by the ratio of the intensities

$$R = \left(\frac{E_{r0}}{E_{i0}}\right)^2 = \left(\frac{\eta_2 - \eta_1}{\eta_2 + \eta_1}\right)^2 \tag{AIV.33}$$

In loss-less media

$$v = \frac{\eta}{\mu} \qquad \text{(AIV.34)}$$

where v is the velocity of propagation and μ is the permeability of the medium. For most cases of interest $\mu \simeq \mu_0$, so that

$$R = \left(\frac{v_2 - v_1}{v_2 + v_1}\right)^2 = \left(\frac{v_2/c - v_1/c}{v_2/c + v_1/c}\right) = \left(\frac{n_2 - n_1}{n_2 + n_1}\right)^2 \qquad \text{(AIV.35)}$$

where n is the refractive index of the medium. Similarly, the *transmissivity*, T, is given by

$$T = \left(\frac{2n_2}{n_2 + n_1}\right)^2 \qquad \text{(AIV.36)}$$

If the incident wave is obliquely incident rather than normally incident, it is necessary to take account of the incidence angle and the angle of the transmitted wave in treating the boundary conditions. Figure AIV.2 illustrates the geometry for one case, that is, for the case where the **E** vector is parallel to the plane of incidence. In this case, using the same notation as for normal incidence,

$$\tau_{\parallel} = \frac{E_{r0}}{E_{i0}} = \frac{\eta_2 \cos \theta_t - \eta_1 \cos \theta_i}{\eta_2 \cos \theta_t + \eta_1 \cos \theta_i} \qquad \text{(AIV.37)}$$

and

$$\Gamma_{\parallel} = \frac{E_{t0}}{E_{i0}} = \frac{2\eta_2 \cos \theta_i}{\eta_2 \cos \theta_t + \eta_1 \cos \theta_i} \qquad \text{(AIV.38)}$$

It is clear from equation (AIV.37) that the reflected **E** wave may vanish if

$$\eta_2 \cos \theta_t = \eta_1 \cos \theta_i \qquad \text{(AIV.39)}$$

The angle of incidence for which this occurs is called the *Brewster angle,* and an arbitrarily polarized wave incident on the surface at this angle will have in its reflected wave only an **E** component parallel to the plane of incidence. Thus an appropriately angled window can serve as a polarizing filter.

If the E field of the incident wave is perpendicular to the plane of incidence, one has

$$\tau_{\perp} = \frac{E_{r0}}{E_{i0}} = \frac{\eta_2 \cos \theta_i - \eta_1 \cos \theta_t}{\eta_2 \cos \theta_i + \eta_1 \cos \theta_t} \qquad \text{(AIV.40)}$$

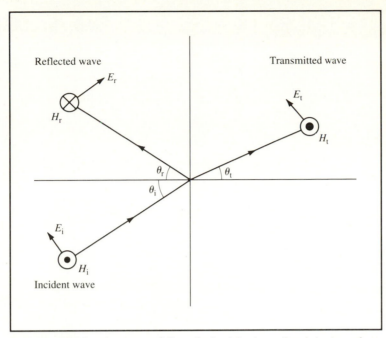

FIGURE AIV.2. A wave obliquely incident on the interface between two media where the electric field is polarized in the plane of incidence.

and

$$\Gamma_{\perp} = \frac{E_{t0}}{E_{i0}} = \frac{2\eta_2 \cos \theta_i}{\eta_2 \cos \theta_t + \eta_1 \cos \theta_i} \qquad \text{(AIV.41)}$$

Equations (AIV.38)–(AIV.41) are known as the *Fresnel equations*, and these equations are the more general equations governing reflection and transmission at an interface. It can be shown that the Brewster angle θ_B for parallel polarization is given by

$$\tan \theta_{B\parallel} = \frac{n_2}{n_1} = \sqrt{\frac{\epsilon_2}{\epsilon_1}} \qquad \text{(AIV.42)}$$

while for polarization perpendicular to the plane of incidence,

$$\sin^2\theta_{B\perp} = \frac{1 - (\mu_1\epsilon_2/\mu_2\epsilon_1)}{1 - (\mu_1/\mu_2)^2} \qquad \text{(AIV.43)}$$

For most media of interest $\mu_1 = \mu_2 = \mu_0$ and $\sin^2 \theta_{B\perp}$ does not exist.

Selected References

General

Born, M., and E. Wolf, *Principles of Optics,* 2nd ed., Pergamon, Oxford, 1964.

Ditchburn, R. W., *Light,* Dover, New York, 1991.

Driscoll, W. G., and W. Vaughn, *Handbook of Optics,* McGraw-Hill, New York, 1978.

Iizuka, K., *Engineering Optics,* Springer-Verlag, New York, 1985.

Nussbaum, A., and R. A. Phillips, *Contemporary Optics for Scientists and Engineers,* Prentice-Hall, Englewood Cliffs, NJ, 1976.

Scott, D. M., *The Physics of Vibrations and Waves,* Merrill, Columbus, OH, 1986.

Optical Design

Kingslake, R., *Lens Design Fundamentals,* Academic Press, New York, 1978.

O'Shea, D., *Elements of Modern Optical Design,* Wiley, New York, 1985.

Smith, W. J., *Modern Optical Engineering,* 2nd ed., McGraw-Hill, New York, 1990.

Fiber Optics

Gowar, J., *Optical Communication Systems,* Prentice-Hall, Englewood Cliffs, NJ, 1984.

Killen, H. B., *Fiber Optic Communications,* Prentice-Hall, Englewood Cliffs, NJ, 1991.

Palais, J. C., *Fiber Optic Communications,* 2nd ed., Prentice-Hall, Englewood Cliffs, NJ, 1988.

Zanger, H., and C. Zanger, *Fiber Optics,* Merrill, New York, 1991.

Fourier Optics

Gaskill, J. D., *Linear Systems, Fourier Transforms, and Optics,* Wiley, New York, 1978.

Goodman, J. W., *Introduction to Fourier Optics,* McGraw-Hill, New York, 1968.

Steward, E. G., *Fourier Optics, An Introduction,* Wiley, New York, 1983.

Sources and Lasers

Lengyel, B. A., *Lasers,* 2nd ed., Wiley, New York, 1971.

O'Shea, D. C., W. R. Callen, and W. T. Rhodes, *An Introduction to Lasers and Their Applications,* Addison-Wesley, Reading, MA, 1977.

Wherrett, B. S., ed., *Laser Advances and Applications,* Wiley, New York, 1980.

Young, M., *Optics and Lasers,* 3rd ed., Springer-Verlag, New York, 1986.

Answers to Selected Problems

Chapter 2

1. 100 m
4. 5.41 h after sundown
6. 131°
12. 1.66
16. $s' = 15$; $\mu = -1.5$

Chapter 3

1. $[\mathbf{S}] = \begin{bmatrix} 1.077129 & -0.029164 \\ 1.786230 & 0.880032 \end{bmatrix}$
7. $d_{cl} = 4.82$
9. (a) $\mu = -0.714$, $s = -42.85$; (b) -5.00, -150; (c) 0.40, 6.00 image virtual and erect; (d) 1.25, 6.25
11. $[\mathbf{S}] = \begin{bmatrix} 1 - \dfrac{2t}{R} & -\dfrac{2n}{R} \\ \dfrac{2t}{n}\left(1 - \dfrac{t}{R}\right) & 1 - \dfrac{2t}{R} \end{bmatrix}$

Chapter 4

1. efl = 91.116173
2. efl = 157.60
3. efl = 10.0146919
5. $\phi_F = -1.448\phi$. Flint element second.
7. $L_{final} = 28.1645$
8. $L_{final} = 88.43646$ marginal trace, object at infinity
    $L_{final} = 162.9408$ zonal trace, object at -200 cm
9. $L_{final} = 156.0163$ zonal trace, object at infinity
    $L_{final} = 731.6295$ marginal trace, object at -200 cm

Chapter 5

1. 1.95×
6. 4 times larger
7. s_o from -195.0 to -205.3 cm
9. $s = 1.59$ cm
11. $\mu_M = 533\times$
15. Exit pupil diameter 6.7 mm

Chapter 6

4. i, ii, v, vii
9. Maxima: $n = 1$, $\theta = 0.315°$; $n = 10$, $\theta = 3.15°$
    Minimum: $n = 5$, $\theta = 1.42°$
11. $\Delta d = 2.275$ nm
15. $\lambda = 579$ nm
19. $I/I_0 = 0.924$
22. $t = 1.587 \ \mu$m

Chapter 7

1. $n = 1.333467$
5. 12,477 slits/mm
10. 44.65 rad/s
12. $R_L = 80.255$ cm
16. 238 nm
17. $\mathcal{F} = 36.73$

Chapter 8

3. $n = 30$
5. $R_p = 2.415$ m
8. $z = 1.36 \times 10^6$ m
13. 571 nm, 444 nm

Chapter 9

1. 1.14
3. 0.113 W
6. 2.12 eV
9. $A = 2.62 \times 10^{-5}$
12. $\lambda_{ZnO} = 387.5$ nm

Chapter 10

1. NA $= 0.9940$
5. $P_{in} = 0.92$ nV
6. 222 μm into the fiber
8. 3.125×10^6 photons
12. 4 links

Chapter 11

5. $X_f = \pm 0.2$ mm for a 500-nm source
9. 5 p/mm
10. $\sin \theta$ between 0.08716 and 0.64278

INDEX